AF352793

Engineering for Surgery

Engineering for Surgery

Special Issue Editors

Nicola Pio Belfiore
Pietro Ursi
Andrea Scorza

MDPI • Basel • Beijing • Wuhan • Barcelona • Belgrade • Manchester • Tokyo • Cluj • Tianjin

Special Issue Editors

Nicola Pio Belfiore
Roma Tre University
Italy

Pietro Ursi
Sapienza University of Rome
Italy

Andrea Scorza
Roma Tre University
Italy

Editorial Office
MDPI
St. Alban-Anlage 66
4052 Basel, Switzerland

This is a reprint of articles from the Special Issue published online in the open access journal *Applied Sciences* (ISSN 2076-3417) (available at: https://www.mdpi.com/journal/applsci/special_issues/Engineering_Surgery).

For citation purposes, cite each article independently as indicated on the article page online and as indicated below:

LastName, A.A.; LastName, B.B.; LastName, C.C. Article Title. *Journal Name* **Year**, *Article Number*, Page Range.

ISBN 978-3-03936-605-7 (Hbk)
ISBN 978-3-03936-606-4 (PDF)

Contents

About the Special Issue Editors

Nicola Pio Belfiore, Full Professor, IEEE Member, teaches Applied Mechanics and Functional Design at the University of Roma Tre, Italy. After being awarded his Ph.D. degree, completed at "Sapienza" in cooperation with the University of Maryland of College Park, he won three international awards in recognition of his research. He has served as Honorary Professor of Obuda University, Hungary, since 2008. The author of three textbooks, three patents, and around one hundred scientific papers, Belfiore has also served as coordinator of numerous scientific projects, both national and European. In 2013, he was Director of the 2nd Level Vocational Master in Energy Conversion Efficiency and Renewable Energy. His current research interests include dynamics, functional design, MEMS, robotics, tribology, and kinematics. In October 2017, he moved from Sapienza to Roma Tre University, where, in December 2020, has was appointed Head of the Degree Programs of Mechanical Engineering, which includes the BSc and MSc in Mechanical Engineering, MSc in Aeronautical Engineering, and BSc and MSc in Marine and Ocean Engineering.

Pietro Ursi is General Surgery University Fixed-Term Researcher at the Department of General and Specialized Surgery "Paride Stefanini" of Sapienza University in Rome, where he teaches courses concerning general surgery at MS level. His activities have been funded through research grants for surgical specialties and organ transplantation. After his achievement of the Italian Laurea Degree in Medicine and Surgery in 2000, Pietro Ursi obtained his Ph.D. in "New Technologies in Surgery" from the Consortium of Marche Polytechnic University and Sapienza University in 2009. He has been working as a specialist in digestive system surgery and digestive endoscopy at Sapienza University and has also been a member of the Italian Medical Association since 2001. He carries out outpatient activities, first aid ward assistance, and surgical operating room activities in small, medium, and large surgeries in addition to serving as a first aid doctor on mobile units.

Andrea Scorza, Researcher, teaches Industrial Measurements at the University of Roma Tre, Italy. He received his Ph.D. degree in Mechanical Measurements for Engineering from the University of Padova, Padua, Italy, in 2004. He has served as Assistant Professor of Measurements and Clinical Engineering at the Department of Engineering, University of Roma Tre, Roma, Italy, since his appointment in 2012. His current research interests include mechanical and thermal measurement systems and instrumentation, design and testing of biomedical instrumentation, and experimental mechanics applied in biomedical fields.

Preface to "Engineering for Surgery"

The interaction between engineering and surgery serves as a source of progress for both sectors. The high standards of surgical operations, in terms demand of accuracy, reliability, and miniaturization, present a challenge for engineers, while some new achievements in engineering, in turn, have inspired numerous surgeons to improve the efficacy and success rate of their operations.

This is an intrinsically vast and interdisciplinary subject and, therefore, the present Special Issue offers only a little sample of the immense variety of applications, some of which have been successfully applied or are still under development, while we have been offered a preview of others thanks to the fantasy of creative science fiction writers.

The purpose of this Special Issue is, therefore, to stimulate the interest of engineers and surgeons, who will benefit from mutual advantages gained from their cooperation. We have been pleased to receive a number of contributions, and we sincerely appreciate all contributions, although the acceptance rate for this Special Issue was around 50%.

Finally, we gratefully acknowledge the entire staff of MDPI for supporting and trusting our work. Our service for Applied Sciences has made us more aware of the activities related to the communication of scientific results. A huge thank you goes to Luca Shao, who encouraged and supported all of us to develop the current Special Issue from the beginning of last year until today.

Nicola Pio Belfiore, Pietro Ursi, Andrea Scorza
Special Issue Editors

Editorial

Engineering-Aided Inventive Surgery

Nicola Pio Belfiore [1,*], **Andrea Scorza** [1] **and Pietro Ursi** [2]

[1]   Department of Engineering, University of Roma Tre, via della Vasca Navale 79, 00146 Rome, Italy; andrea.scorza@uniroma3.it

[2]   Department of General Surgery and Surgical Specialties "Paride Stefanini", Sapienza University of Rome, Viale del Policlinico 155, 00161 Rome, Italy; pietro.ursi@uniroma1.it

*   Correspondence: nicolapio.belfiore@uniroma3.it; Tel.: +39-06-5733-3316

Received: 1 June 2020; Accepted: 3 June 2020; Published: 7 June 2020

Abstract: This Editorial presents a new Special Issue dedicated to some old and new interdisciplinary areas of cooperation between engineering and surgery. The first two sections offer some food for thought, in terms of a brief introductory and general review of the past, present, future and visionary perspectives of the synergy between engineering and surgery. The last section presents a very short and reasoned review of the contributions that have been included in the present Special Issue. Given the vastness of the topic that this Special Issue deals with, we hope that our effort may have offered a stimulus, albeit small, to the development of cooperation between engineering and surgery.

Keywords: engineering; surgery; interdisciplinarity

---

1. Introduction

For centuries, humankind has been dreaming about how to save lives and pursue immortality, and for this reason medicine and surgery have been always fundamental topics. Fiction and science fiction illustrate clearly the vast variety of expectations that people have from progress in these two important fields. An example of how human vision pushes forward the most secret ambitions is described in *Frankenstein* Mary Shelley's 1818 novel [1], wherein the doctor main character puts together pieces of dead bodies to build a new body and makes him alive with electricity. In the American epic space-opera *Star Wars* (Lucasfilm, 1977) there are also many examples of how medicine is expected to be in a far future: the prosthetic hand that replaces Luke's lost one, in a perfectly equivalent manner; the hibernation techniques; the robotic obstetricians; the fully automated orthopedics apparatus that allows a total replacement of the lower limbs; and Darth Vader's portable automatic respirator. Another interesting example of futuristic surgery has been suggested in *Fantastic Voyage* (20th Century Fox, 1966), which nowadays receives an exaggerated number of mentions at conferences by authors presenting their work in micro or nanosurgery. According to the plot, after an incredible miniaturization process, a submarine about the size of a microbe flows in a patient's ducts to remove a blood clot in his brain. Additionally, the task of producing a correct and fast diagnosis is every doctor's secret dream. In the science fictional series *Star trek* (Desilu Productions, 1966), chief medical officer Dr. McCoy obtains an immediate and detailed diagnosis simply moving a small sensor back and forth over the patient.

Unfortunately, we are far enough away from these goals, and therefore hard, maybe impossible, work remains for us to do along the road ahead. One way to start our endeavor consists of straightening the cooperation between medicine end engineering, because any progress in any technical apparatus that gives enhancements in surgery is based on both the fields of application. Furthermore, a great amount of creativity and interdisciplinary approach is needed to enhance the developments of new tools, which we could refer to as *inventive engineering for surgery* or *engineering-aided inventive surgery*.

The need for cooperation is intrinsically related to the fact that doctors *know what to do*, while engineers *know how to help to make it real* or may even suggest new facilities that open new scenarios for unheard-of operations. At first, progress may have been developed thanks to the creativity of surgeons who had a bit of technical know-how or who dared to experiment with new technical stuff. Again, making reference to another American television drama series, in *The Knick* (AMBEG Screen Products, 2014), Dr. Thackery, chief surgeon at the Knickerbocker Hospital, in case of emergencies, enhances surgical procedures by using the technical equipment in very creative arrangements.

After all these fictional examples, it is a pleasure to mention the exciting and enlightening survey of real cases in surgery recently written by van de Laar [2], where 28 *historical* operations have been described in terms so clear that even an engineer can understand. Among the conclusive remarks, it is very agreeable to assess that, for the moment, *there is as yet no question of computers completely taking over the tasks of human doctors*. However, the described operations show how the cooperation between engineering and medicine has been or could have been important to complete the task with success.

2. From Early Tools to the State of the Art

One way to explain how the cooperation between Technology and Surgery works consists in interpreting it as a customer-provider relationship where engineering offers new technological developments to the surgery' demand. In order to appreciate how strong the surgery–engineering relationship is, let us consider the following classical and fundamental topics in engineering and some of their representative applications:

- Design, strength of materials, and material development, which are required to develop any form of surgical tool;
- Kinematics and dynamics that are necessary to build non-stationary systems;
- Measurements and control that are required in the operational environment;
- Nanotechnology, microelectronics, information technology, and telecommunications that are necessary to develop the operational equipment;
- Pneumatic and fluid dynamics that are fundamental to sustain the vital function of the patient during operation.

Many other branches of engineering are relevant too. All of these capabilities are also necessary to develop most of the new frontiers of surgery, such as

- Smart surgical tools;
- Micro and nano robots for surgery;
- Minimally invasive procedures.

They can be applied to general, lung, gynecologic, head and neck, heart, neuro-spine, vascular, and urological surgery.

Classical Fields of Application

The above-mentioned relationship between engineering and surgery cannot be described in terms of a simple offer-demand interchange. In fact, this cooperation is quite complex and multifaceted a liaison: any possible development in surgery could be supported by a proper collaboration with the engineering counterparts, while almost any new development in engineering could be successfully applied to improve surgical operations. Both alliances need a strong and very integrated partnership and an enduring team work. For all these reasons, the original call for papers from this Special Issue has been open to the following general topics:

- Laparoscopic surgery [3,4];
- Endoscopic surgery [5];
- Robotic surgery [6];
- Natural orifice transluminal endoscopic surgery (NOTES) [7,8];
- New technologies for intraoperative imaging [9].

While more specific topics have been also solicited:

- New technologies for training of residence and young surgeons in minimally invasive surgery [10];
- New technologies for the development of MEMS/NEMS and microsystems for surgery, such as topological [11–13] kinematic synthesis, smart fabrication of multi-DoF crawling tools [14] and operational capability [15–17];
- Transanal endoscopic microsurgery (TEM) [18] and transanal minimally invasive surgery (TAMIS) [19];
- Ethics: ethical issues in the application of autonomous robots in surgery [20];
- Education: new trends in teaching–learning methods and information technology [21].

The next section offers a very short and reasoned review of the contributions that have been included in the present Special Issue. Given the vastness of the topic that this Special Issue deals with, we hope that our effort may have offered a stimulus, albeit small, to the development of cooperation between engineering and surgery.

3. About the Present Issue

Nowadays the science of engineering may support surgery in different ways and through synergies that were hardly conceivable until a few years ago, thanks to the recent advances in applied sciences and technology. In particular, engineering contributions may range from pre-operative assessment to post operative care, and from computer aided-surgery to hardware development for performance improvement of consolidated treatments or novel surgical approaches, such as management and characterization of surgical tools and instrumentation.

Therefore, in this special issue some stimulating contributions are proposed for their valuable applications into the pre-operative field, focusing on modern simulation methods and 3D imaging tools for surgical planning, prediction methodologies [22,23] and data acquisition by means of novel wearable devices [24].

Anyway, the operating field in the context of synergies between engineering and surgery provides most of the advanced and promising solutions. In this regard, further interesting applications are described in the issue: from the control of MRI-compatible robots [25] and the guidance of surgical needles [26], to the use of very complex image analysis methods for surgical tool characterization [27] and the development of novel devices with high functional performances [28] and better ergonomic design for laparoscopic applications [29].

Moreover, solutions and innovations become very forward-thinking when engineering science is challenged with the requirements of minimally invasive surgery, as reported, for example, in the paper [30] where cataract surgery is combined with high frequency deep sclerotomy (HFDS). One more article [31] concerns the novel laser assisted in situ Keratomileusis (LASIK) applications for vision correction in myopia and presbyopia diseases.

Finally, a significant part of modern surgery relies on pioneering efforts to bring about advances in nano and microengineering to the lab and surgical activities. In this issue, an example of extreme miniaturization of the tools used in surgery has been provided by two papers, one concerning the development of a new piezo MEMS tweezer for soft materials characterization [32] and one describing some reasonable progress of MEMS for operations in a surgical scenario [33].

Funding: This research received no external funding.

Conflicts of Interest: The authors declare no conflict of interest.

References

1. Wollstonecraft Shelley, M. *Frankenstein, or, the Modern Prometheus: The 1818 Text*; Oxford University Press: Oxford, MI, USA; New York, NY, USA, 1998.

2.   Van de Laar, A. *Under the Knife: A History of Surgery in 28 Remarkable Operations*, English Edition; John Murray: London, UK, 2018.

3.   Neudecker, J.; Sauerland, S.; Neugebauer, E.; Bergamaschi, R.; Bonjer, H.J.; Cuschieri, A.; Fuchs, K.H.; Jacobi, C.; Jansen, F.W.; Koivusalo, A.M.; et al. The European Association for Endoscopic Surgery clinical practice guideline on the pneumoperitoneum for laparoscopic surgery. *Surg. Endosc.* **2002**, *16*, 1121–1143. [CrossRef] [PubMed]

4.   Balla, A.; Quaresima, S.; Ursi, P.; Seitaj, A.; Palmieri, L.; Badiali, D.; Paganini, A.M. Hiatoplasty with crura buttressing versus hiatoplasty alone during laparoscopic sleeve gastrectomy. *Gastroenterol. Res. Pract.* **2017**, *2017*, 6565403. [CrossRef] [PubMed]

5.   Rattner, D.; Kalloo, A. ASGE/SAGES Working Group on Natural Orifice Translumenal Endoscopic Surgery: October 2005. *Surg. Endosc.* **2006**, *20*, 329–333. [CrossRef] [PubMed]

6.   Lanfranco, A.R.; Castellanos, A.E.; Desai, J.P.; Meyers, W.C. Robotic Surgery: A Current Perspective. *Ann. Surg.* **2004**, *239*, 14–21. [CrossRef]

7.   Bardaro, S.; Swanström, L. Development of advanced endoscopes for natural orifice transluminal endoscopic surgery (NOTES). *Minim. Invasive Ther. Allied Technol.* **2006**, *15*, 378–383. [CrossRef]

8.   Quaresima, S.; Paganini, A.M.; D'Ambrosio, G.; Ursi, P.; Balla, A.; Lezoche, E. A modified sentinel lymph node technique combined with endoluminal loco-regional resection for the treatment of rectal tumours: A 14-year experience. *Color. Dis.* **2017**, *19*, 1100–1107. [CrossRef]

9.   Nimsky, C.; Ganslandt, O.; Von Keller, B.; Romstöck, J.; Fahlbusch, R. Intraoperative high-field-strength MR imaging: Implementation and experience in 200 patients. *Radiology* **2004**, *233*, 67–78. [CrossRef]

10.   Van Der Meijden, O.A.J.; Schijven, M.P. The value of haptic feedback in conventional and robot-assisted minimal invasive surgery and virtual reality training: A current review. *Surg. Endosc.* **2009**, *23*, 1180–1190. [CrossRef]

11.   Liberati, A.; Belfiore, N.P. A method for the identification of the connectivity in multi-loop kinematic chains: Analysis of chains with total and partial mobility. *Mech. Mach. Theory* **2006**, *41*, 1443–1466. [CrossRef]

12.   Belfiore, N.P. Distributed Databases for the development of Mechanisms Topology. *Mech. Mach. Theory* **2000**, *35*, 1727–1744. [CrossRef]

13.   Pennestrì, E.; Belfiore, N.P. On the numerical computation of Generalized Burmester Points. *Meccanica* **1995**, *30*, 147–153. [CrossRef]

14.   Verotti, M.; Bagolini, A.; Bellutti, P.; Belfiore, N.P. Design and validation of a single-SOI-wafer 4-DOF crawling microgripper. *Micromachines* **2019**, *10*, 376. [CrossRef] [PubMed]

15.   Crescenzi, R.; Balucani, M.; Belfiore, N.P. Operational characterization of CSFH MEMS technology based hinges. *J. Micromech. Microeng.* **2018**, *28*, 055012. [CrossRef]

16.   Potrich, C.; Lunelli, L.; Bagolini, A.; Bellutti, P.; Pederzolli, C.; Verotti, M.; Belfiore, N.P. Innovative silicon microgrippers for biomedical applications: Design, mechanical simulation and evaluation of protein fouling. *Actuators* **2018**, *7*, 12. [CrossRef]

17.   Vurchio, F.; Ursi, P.; Buzzin, A.; Veroli, A.; Scorza, A.; Verotti, M.; Sciuto, S.A.; Belfiore, N.P. Grasping and releasing agarose micro beads in water drops. *Micromachines* **2019**, *10*, 436. [CrossRef]

18.   Middleton, P.F.; Sutherland, L.M.; Maddern, G.J. Transanal endoscopic microsurgery: A systematic review. *Dis. Colon Rectum* **2005**, *48*, 270–284. [CrossRef]

19.   Hong, K.D.; Kang, S.; Urn, J.W.; Lee, S.I. Transanal Minimally Invasive Surgery (TAMIS) for Rectal Lesions: A Systematic Review. *Hepato Gastroenterol.* **2015**, *62*, 863–867.

20.   Satava, R.M. Laparoscopic Surgery, Robots, and Surgical Simulation: Moral and Ethical Issues. *Surg. Innov.* **2002**, *9*, 230–238. [CrossRef]

21.   Turner, S.R.; Mormando, J.; Park, B.J.; Huang, J. Attitudes of robotic surgery educators and learners: Challenges, advantages, tips and tricks of teaching and learning robotic surgery. *J. Robot. Surg.* **2020**, *14*, 455–461. [CrossRef]

22.   Olivetti, E.C.; Nicotera, S.; Marcolin, F.; Vezzetti, E.; Sotong, J.; Zavattero, E.; Ramieri, G. 3D Soft-tissue prediction methodologies for orthognathic surgery—A literature review. *Appl. Sci.* **2019**, *9*, 4550. [CrossRef]

23.   Wang, J.; Yang, Y.; Guo, D.; Wang, S.; Fu, L.; Li, Y. The effect of patellar tendon release on the characteristics of patellofemoral joint squat movement: A simulation analysis. *Appl. Sci.* **2019**, *9*, 4301. [CrossRef]

24.   Pajic, B.; Zakharov, P.; Pajic-Eggspuehler, B.; Cvejic, Z. User friendliness of awearable visual behavior monitor for cataract and refractive surgery. *Appl. Sci.* **2020**, *10*, 2190. [CrossRef]

25. Meng, D.; Lu, B.; Li, A.; Yin, J.; Li, Q. Pressure observer based adaptive dynamic surface control of pneumatic actuator with long transmission lines. *Appl. Sci.* **2019**, *9*, 3621. [CrossRef]

26. Li, Z.; Song, S.; Liu, L.; Meng, M.Q.H. Tip estimation method in phantoms for curved needle using 2D transverse ultrasound images. *Appl. Sci.* **2019**, *9*, 5305. [CrossRef]

27. Scarano, A.; Noumbissi, S.; Gupta, S.; Inchingolo, F.; Stilla, P.; Lorusso, F. Scanning Electron Microscopy Analysis and Energy Dispersion X-ray Microanalysis to Evaluate the Effects of Decontamination Chemicals and Heat Sterilization on Implant Surgical Drills: Zirconia vs. Steel. *Appl. Sci.* **2019**, *9*, 2837. [CrossRef]

28. Belfiore, N.P. A new concept compliant platform with spatial mobility and remote actuation. *Appl. Sci.* **2019**, *9*, 3966. [CrossRef]

29. Sánchez-Margallo, J.A.; González, A.G.; Moruno, L.G.; Gómez-Blanco, J.C.; Pagador, J.B.; Sánchez-Margallo, F.M. Comparative study of the use of different sizes of an ergonomic instrument handle for laparoscopic surgery. *Appl. Sci.* **2020**, *10*, 1526. [CrossRef]

30. Pajic, B.; Cvejic, Z.; Mansouri, K.; Resan, M.; Allemann, R. High-frequency deep sclerotomy, a minimal invasive Ab interno glaucoma procedure combined with cataract surgery: Physical properties and clinical outcome. *Appl. Sci.* **2020**, *10*, 218. [CrossRef]

31. Pajic, B.; Cvejic, Z.; Massa, H.; Pajic-Eggspuehler, B.; Resan, M.; Studer, H.P. Laser vision correction for regular myopia and supracor presbyopia: A comparison study. *Appl. Sci.* **2020**, *10*, 873. [CrossRef]

32. Botta, F.; Rossi, A.; Belfiore, N.P. A feasibility study of a novel piezo MEMS tweezer for soft materials characterization. *Appl. Sci.* **2019**, *9*, 2277. [CrossRef]

33. Vurchio, F.; Ursi, P.; Orsini, F.; Scorza, A.; Crescenzi, R.; Sciuto, S.; Belfiore, N.P. Toward operations in a surgical scenario: Characterization of a microgripper via light microscopy approach. *Appl. Sci.* **2019**, *9*, 1901. [CrossRef]

 *applied sciences*

Review

3D Soft-Tissue Prediction Methodologies for Orthognathic Surgery—A Literature Review

Elena Carlotta Olivetti [1], Sara Nicotera [1], Federica Marcolin [1,*], Enrico Vezzetti [1], Jacqueline P. A. Sotong [2], Emanuele Zavattero [2] and Guglielmo Ramieri [2]

[1] Department of Management and Production Engineering, Politecnico di Torino, corso Duca degli Abruzzi 24, 10129 Torino, Italy; elena.olivetti@polito.it (E.C.O.); sara.nicotera@studenti.polito.it (S.N.); enrico.vezzetti@polito.it (E.V.)

[2] Department of Maxillofacial unit, Città della Salute e della Scienza HOSPITAL, 10129 Torino, Italy; slim_js2000@yahoo.fr (J.P.A.S.); emanuele.zavattero@gmail.com (E.Z.); guglielmo.ramieri@unito.it (G.R.)

* Correspondence: federica.marcolin@polito.it

Received: 2 July 2019; Accepted: 22 October 2019; Published: 26 October 2019

Abstract: Three-dimensional technologies have had a wide diffusion in several fields of application throughout the last decades; medicine is no exception and the interest in their introduction in clinical applications has grown with the refinement of such technologies. We focus on the application of 3D methodologies in maxillofacial surgery, where they can give concrete support in surgical planning and in the prediction of involuntary facial soft-tissue changes after planned bony repositioning. The purpose of this literature review is to offer a panorama of the existing prediction methods and software with a comparison of their reliability and to propose a series of still pending issues. Various software are available for surgical planning and for the prediction of tissue displacements, but their reliability is still an unknown variable in respect of the accuracy needed by surgeons. Maxilim, Dolphin and other common planning software provide a realistic result, but with some inaccuracies in specific areas of the face; it also is not totally clear how the prediction is obtained by the software and what is the theoretical model they are based on.

Keywords: orthognathic surgery; 3D face analysis; surgical planning; soft tissue prediction; prediction methods

---

1. Introduction

Patients presenting dentofacial deformities are commonly subject to combined orthodontic and surgical treatment such as Le Fort I osteotomy (LFI), bilateral sagittal split osteotomy (BSSO), intraoral ramus vertical osteotomy (IVRO), sagittal split ramus osteotomy (SSRO), bimaxillary surgery and genioplasty. These interventions have been commonly planned with two-dimensional methodologies; today, the last challenge is the three-dimensional (3D) surgery planning. The refining of 3D graphics and imaging tools gives the chance to explore surgical planning and prediction of the effects of different clinical approaches; these techniques are based on images acquired with computed tomography (CT), cone bean computed tomography (CBCT) and multi-slice computed tomography (MSCT), which provide volumetric images of facial anatomical structure. Additionally, the surface of the face can be scanned and mapped to underline the effect of changes in facial appearance using 3D laser technology [1] that gives a great contribution to surgeons to decide on the type of surgeries as well as on the magnitude and direction of surgical movements to correct facial dysmorphology. Moreover, a deep interest in the prediction of soft-tissue response to hard tissue movements has been growing and two-dimensional conventional methodologies seem to be insufficient for this aim as they do not take into account the third dimension.

Indeed, the possibility to know the soft-tissue response to surgical operations helps surgeons to plan surgical movements and it gives surgeons more information about the need of orthodontic decompensation. Furthermore, the purpose of these interventions is not only to correct the facial dysmorphology from a functional point of view, but also to obtain an aesthetic enhancement of patients' facial aspect. Therefore, an accurate treatment planning is very important to obtain a good aesthetic and occlusal result [2]. For the aforementioned reasons, having a preview of the soft-tissue arrangement is extremely important. In future, this opportunity could bring to have a surgical planning methodology able to best match patients' aesthetic expectations, additionally to the functional corrective aspect.

Several methods have been considered to forecast soft tissue responses; the most common are the mass-spring model (MSM), finite element model (FEM) and mass tensor model (MTM). Most of the software packages currently adopted in clinical practice are based on these models. Generally, those softwares seem to reach an acceptable overall accuracy, but with inaccuracies for specific areas of the face, for example around the lips.

Studies have been conducted to develop 2D prevision models based on the ratio between facial and bony movements, particularly focusing on face sub regions [3]; by comparing predictive models constructed on this ratio and patients' post-operative conditions, it is stated that traditional approaches show limits in forecasting the outcome of large and complex movements. Moreover, a large number of variables must be selected [4].

Although there are several studies on the soft tissue changes after maxillary osteotomies, few of them report a systematic analysis.

The target of this systematic review is to gather information on the existing prediction methods and software of soft tissue displacements after dysmorphism corrective surgery, and to draw conclusions on the accuracy and reliability of these software in the preview of surgical outcome. Additionally, some problems related to the level of accuracy needed by surgeons, to the predictive imprecision reported in the studies and to the magnitude of the acceptable error are presented.

This work is structured as follows. The section Material and Methods describes the methods that have been adopted for the literature review. The section Results includes the clinical details of patients involved and the details of the articles considered; the articles have been divided in subsections according to the software used by the authors. Finally, the section Discussion and Conclusions discusses the current prediction methods and concludes the work.

2. Materials and Methods

The research of this systematic review is based on the Population Intervention Control Outcome Study design (PICOS) format (Table 1). PubMed and Scopus are the databases adopted for our research. The considered keywords are: orthognathic surgery, facial dysmorphism, surgical planning, 3D, 3D face analysis, soft tissue, BSSO, bilateral sagittal split osteotomy, Le Fort I, prediction methods.

The research has been set on different combinations of keywords; at first, we focused generically on the orthognathic surgery, then the type of interventions has been specified: BSSO, IVRO, SSRO and LFI. Finally, the research has been limited to soft tissue simulation and prediction.

All the articles found were assessed by three authors and classified in: prospective study (PS), retrospective study (RS), case series study (CS).

Single cases reports have been excluded from our analysis because they have been considered not clinically significant.

Articles published before 2000 have not been considered.

The following data have been recorded for each eligible study: first author and year of publication, journal, study design, sample size, gender, mean age, diagnosis, imaging technology, typologies of surgery, software used for the prediction, soft and hard tissue landmarks considered, time interval from surgery to post-surgical imaging, results and conclusions of the authors.

Table 1. PICOS (Population Intervention Control Outcome Study) criteria for the systematic review.

Population	Patients with angle class II, III dentoskeletal deformities, indicated for a maxillary osteotomy to correct the malocclusion
Intervention	Le Fort I osteotomy, bimaxillary osteotomy, BSSO (bilateral sagittal split osteotomy), IVRO (intraoral ramus vertical osteotomy), SSRO (sagittal split ramus osteotomy), genioplasty
Comparison	3D orthognathic surgery planning and prediction method
Outcome	Soft tissue post-operative change, prediction and accuracy of the technique
Study design	Clinical trials, retrospective and prospective studies (CT, RS and PS, respectively) with the aim of assessing the methodologies of soft and hard tissue prediction.

3. Results

Different responses have been obtained varying the insertion order of the keywords. After having combined the outcomes and removed the duplicates, the remaining articles have been evaluated on the basis of their relevance to the topic. In the end, 24 articles to be deepened and a number of interesting articles to be referenced for a better comprehension of the topic have been selected.

3.1. Clinincal Details

A total of 24 articles have been compared in this work; 12 are retrospective studies, two are case studies and five are prospective studies. Remaining articles have not declared categorization. The sample of patients involved in each study vary from seven to 100 subjects. The soft tissue prediction has been assessed for Le Fort I osteotomy, BSSO, BSSRO and genioplasty in the correction of different types of facial dysmorphism.

Tables 2 and 3 report a brief summary of the referenced papers we focused on. In particular, the demographic details of the patients are summarized in Table 2. A brief overview of methodologies and results of each article is reported in Table 3.

CBCT has been used for the assessment of soft tissues changes in twelve of the twenty-four studies, with the addition of 3D photographs for one article, while cephalometric radiographs have been used in the remaining eight works, with the addition of CT and MSCT for two of them.

The timing of post-surgical imaging has been stated in all articles.

3.2. Prediction Methodologies

Several approaches have been considered to make a mathematical three-dimensional prediction of soft tissue changes after orthognathic surgery. MSM, FEM [5,6] and MTM are the most common. These have been developed into software packages which are currently used in clinical practice. The functioning of these software is generally acceptable if we consider the creation of a plausible facial outcome, but this prediction does not necessarily match with the real outcome. Moreover, studies show that the prediction accuracy decreases for specific facial areas, especially around the lip, and with the increasing complexity of the surgery (larger bony repositioning often results in a higher inaccuracy of the prediction). The studies presented in this literature review evaluate soft-tissue predictions obtained with different software packages.

Table 2. Demographic details of patients (acronyms: PS is prospective study, RS is retrospective study, CS is case series study, NA is not available/not reported).

References	Journal	Study Design	Sample Size	Male Pts.	Female Pts.	Mean Age	Diagnosis	Surgery
N. Abe et al. 2015	Int J Oral and maxillofacial surgery	RS	45	NA	NA	NA	Asymmetry, Class I, Class II	LFI, SSRO, IVRO
C.M. Resnick et al. 2016	J oral maxillofacial surgery	RS	7	2	5	18.1	Maxilla hypoplasia	A single segment LFI osteotomy
A. Bianchi et al. 2010	J oral maxillofacial surgery	NA	10	6	4	24	NA	LFI, BSSO, genioplasty
B. Khambay et al. 2015	Int J Oral and maxillofacial surgery	NA	10	NA	NA	NA	NA	LFI
M.I. Shafi et al. 2013	Int J Oral and maxillofacial surgery	NA	13	8	5	NA	NA	LFI
J. Liebregts et al. 2014	J of cranio-maxillo-facial surgery	NA	60	15	45	26	NA	LFI, BSSO
A. Marchetti et al. 2011	Int J Oral and maxillofacial surgery	NA	10	5	5	NA	NA	LFI, BSSO, genioplasty
T. Mundluru et al. 2017	Int J Oral and maxillofacial surgery	RS	13	NA	NA	NA	Asymmetry	Bimaxillary, BSSO
Ana de Lourdes sa de lira et al. 2012	J Oral Maxillofacial Surgery	CS, RS	80	19(group1) 20(group2)	21(group1) 20(group2)	22.8(group1) 22.4(group2)	Class II	group1: BSSO group2: bimaxillary
Ana de Lourdes sa de lira et al. 2012	J Oral Maxillofacial Surgery	CS, RS	76	10(group1) 26(group2)	26(group1) 14(group2)	25.4(group1) 25.6(group2)	Class III	LFI, bimaxillary
A. Terzic et al. 2013	Aesthetic Plastic Surgery	RS	13	5	8	25.2	NA	Bimaxillary (10 pts.) BSSO (3 pts.)
R.J. Peterman et al. 2016	Progress in Orthodontic	RS	14	3	11	22.5	Class III	LFI, BSSO
R. Ullah et al. 2014	British J of Oral and Maxillofacial Surgery	RS	13	5	8	23	Class III	LFI
Osvaldo Magro Filho et al. 2010	Am J Orthod Dentofacial Orthop	NA	10	NA	NA	NA	Class III	LFI, BSSO
D. Holzinger et al. 2018	J of cranio-maxillo-facial surgery	PS	16	8	8	26	Class III,II,I	NA
Chay Hui Koh et al. 2004	J Oral Maxillofacial Surgery	NA	35	14	21	22.8	Class III	Bimaxillary
O. Donatsky et al. 2009	J of cranio-maxillo-facial surgery	PS	52	21	31	20	Class III	LFI, SSRO
Z.Ö. Pektas et al. 2007	Int J Med and Robotics Comput Assist Surg	RS	11	4	7	23.5	NA	LFI, BSSO
De Riu et al. 2018	J of cranio-maxillo-facial surgery	RS	49	19	30	26.4	NA	Bimaxillary
Van Hemelen et al. 2015	J of cranio-maxillo-facial surgery	PS	66	29	37	NA	Angle class II/class III malocclusion	Bimaxillary, BSSO, LFI, genioplasty
Knoops et al. 2019	Int J Oral and maxillofacial surgery	RS	7	2	5	18	Maxillary sagittal hypoplasia	LFI maxillary advancement
Liebregts et al. 2015	J Oral Maxillofacial Surgery	PS	100	35	65	32	Nonsyndromic mandibular hypoplasia	BSSO
Nam et al. 2015	J of craniofacial surgery	PS	29	16	13	NA	NA	LFI, BSSRO, genioplasry
Nadjmi et al. 2013	Am J Orthod Dent Orthop	RS	13	2	11	NA	NA	LFI, BSSO

Table 3. Imaging technology, software used for prediction, number of landmarks for soft tissue and hard tissue, time interval from surgery to post-operative imaging and results.

Reference	Imaging Technology	Software Adopted for Prediction	Adoption of Land-Marks	Soft Tissue Land-Marks	Hard Tissue Land-Marks	Time Interval from Surgery to Post-Surgery CT	Results	Conclusions
N. Abe et al. 2015	CBCT	OrthoForecast	yes	11 on frontal image, 13 on lateral image, 3 lines	None	12 months	Mean difference in the location of points on predicted and actual postoperative <3.4 mm and <10° in considered angles	Useful, accurate, reliable software to predict soft tissue changes after surgery
C.M. Resnick et al. 2016	CBCT, 3D photographs	Dolphin 3D	yes	37	42	14 months	Mean linear prediction error of 2.91 ± 2.16 mm; mean error at nasolabial angle of 8.1° ± 5.6°. Poorest agreement at subnasale. Most accurate lateral point: maxillary buttress (mean error 3.31 ± 1.81 mm). Least accurate lateral point: lateral ala (mean error 4.54 ± 1.97 mm)	Prediction accuracy acceptable for linear changes in the midline, not for lateral points. In the midline changes at nasal base most prone to prediction error (mean error 2.84 ± 2.55 mm)
B. Khambay et al. 2015	CBCT	3dMDvul-tus	yes	10	None	6 months	85.2%–94.4% points with error <2 mm; RMS error 0.94–2.49 mm	The use of specific anatomic regions is more meaningful than full face
M.I. Shafi et al. 2013	CBCT	Maxilim	yes	NA	NA	6–12 months	Accuracy <3 mm except for upper lip region	Clinically satisfactory for assessed regions but associated with marked errors in upper lip region
J. Liebregts et al. 2014	CBCT	Maxilim	yes	11	None	12 months	Mean absolute error for whole face: 0.81 ± 0.22 mm. Accuracy for whole face, upper lip, lower lip, chin subregion: 100%, 93%, 90%, 95% respectively.	The prediction accuracy is influenced by the magnitude of maxillary/mandibular advancement, age of patient, usage of V-Y closure. Low predictability on upper/lower lip regions.
A. Marchetti et al. 2011	MSCT, cephalo-metric radio-graphs	SurgiCase_CMF	yes	NA	NA	6 months	Mean abs error: 0.50–1.15 mm; percentage of simulation with error ≤ 2 mm: 76–99%	Reliability of soft tissue preview > 91% Realistic, accurate forecast of pt's face after surgery
D. Holzinger et al. 2018	CT, cephalo-metric radio-graphs	Sotirios planning software	yes	NA	15	6 months	Mean error of 1.46 mm ± 1.53 mm	Quite accurate software; clinically satisfactory 3D soft tissue prediction
A. Bianchi et al. 2010	CBCT	SurgiCase_CMF	yes	NA	NA	6 months	Average abs error: 0.94 mm. St.dev: 0.90 mm Percentage of error <2 mm for 86.8%	Reliability results shows that the software makes a realistic preview with low X-ray exposure
T. Mundluru et al. 2017	CBCT	Maxilim	yes	NA	NA	6–12 months	Mean abs distance < 2 mm for all regions.	Prediction of soft tissue clinically acceptable; prediction error < 2 mm; under-prediction around the cheek and chin medio-laterally with opposite effect at inferior border of mandible bilaterally
O. Donatsky et al. 2009	Cephalometric radiographs	TIOPS	yes	8	8	5–6 weeks	Mean accuracy: 0.0–0.5 mm. Inaccuracies: 0.2–1.1 mm	Useful in surgical simulation, planning and prediction. High variability of predicted individual outcome for hard and soft tissue

Table 3. *Cont.*

Z.Ö. Pektas et al. 2007	Cephalometric radiographs	Dolphin imaging version 10.0	yes	7	5	12 months	Least accurate site: lower lip. Most accurate sites: tip of nose, subnasale. Mean difference < 1 mm (4 pts on 7)	Satisfactory in predicting soft tissue outcome
R.J. Peterman et al. 2016	Lateral cephalometric radiographs	Dolphin imaging version 11.0.03.37	yes	9	14	6 months	Error range: ±2 mm (X axis).	Not useful for precise treatment planning
Ana de Lourdes sa de Lira et al. 2012	Lateral cephalogram	Dolphin imaging	no	NA	NA	12 months	Surgeries more extensive than planned. Similar values between predicted and actual results for facial convexity and distance from lips to cranial base	Statistical differences do not invalidate surgical predictability because predicted modifications close to surgical results
Ana de Lourdes sa de Lira et al. 2012	Lateral cephalogram	Dolphin imaging version 11.0	no	NA	NA	12 months	Predicted surgical reduction in mandibular length and angle significantly greater than actual post-operative	Statistical differences do not invalidate surgical predictability because predicted modifications close to surgical results
R. Ullah et al. 2014	CBCT	3dMDvultus	yes	NA	NA	6–12 months	Mean abs distance between actual and predicted soft tissue < 3 mm and ranged from 0.65 mm (chin) to 1.17 (upper lip)	3D soft tissue prediction is acceptable for LFI advancement osteotomy
Osvaldo Magro Filho et al. 2010	Lateral cephalometric radiographs	Dentofacial planner plus (DFP), Dolphin imaging (DI)	no	NA	NA	6 months	No statistical differences between the two software ($p = 0.7945$)	DFP better in predicting nasolabial angle, upper/lower lips. DI better in predicting nasal tip, chin, sub mandibular area
A. Terzic et al. 2013	CT, CBCT	3dMDvultus version 2.2.0.8	yes	NA	NA	6 months	Mean difference between surfaces: −0.64 mm (surgically treated lower half). Errors >3 mm: 4% for upper half and 29.8% for lower half	Accuracy in soft tissue prediction is insufficient
Chay Hui Koh et al. 2004	Lateral cephalograms	CASSOS	yes	16	NA	6 months	Significant differences between 16 of 32 soft tissue measurements ($p < 0.05$) Under estimation of vertical position for upper/lower lips, over estimation of horizontal position of lower lip	Clinically useful prediction of soft tissue profile change in Chinese patients
De Riu et al. 2018	Frontal/lateral cephalometric radiograph, CBCT	Maxilim, Dolphin	yes	NA	NA	3–5 days	Differences among 12 of the 15 parameters were considered not significant. Significant differences reported for SNA ($p = 0.008$), SNB ($p = 0.006$), anterior facial height ($p = 0.033$). Average error: 1.98 mm (linear measures), 1.19° (angle measures)	Good degree of accuracy for most of parameters assessed. Tendency towards under-projection of jaws. 3D simulation of soft tissue currently does not allow an accurate management of facial height and chin position
Van Hemelen et al. 2015	Frontal/lateral cephalometric radiograph, CBCT	Onyx Ceph, Maxilim	yes	17 (3D) 16 (2D)	17 (3D) 16 (2D)	4 months	horizontal direction: 2.29 mm (2D) vs. 1.48 mm (3D) vertical direction: 2.07 mm (2D) vs. 1.46 mm (3D)	3D method more predictive, but more time consuming and more difficult to use.

Table 3. *Cont.*

Knoops et al. 2019	CBCT	Dolphin 3D, ProPlan CMF, PFEM	Yes (Dolphin 3D)	NA	NA	1 year	Dolphin 3D: RMS = 1.8 ± 0.8 mm ProPlan: RMS = 1.2 ± 0.4 mm PFEM: RMS = 1.3 ± 0.4 mm	Dolphin 3D: limited accuracy ProPlan/PFEM: better 3D prediction, mismatch in maxillary position (predicted/real)
Liebregts et al. 2015	CBCT	Maxilim	yes	5	5	6 months	Mean abs. err: whole face: 0.9 ± 0.3 mm upper lip: 1.2 ± 0.5 mm chin area: 0.8 ± 0.5 mm	Accurate soft tissue simulation for single surgery
Nam et al. 2015	CT	Simplant Pro	yes	NA	30	6 months	Mean differences: 0.73 mm (x axis), 1.39 mm (y axis), 0.85 mm (z axis). Accuracy: 52.8%	It is needed to improve soft tissue prediction accuracy
Nadjmi et al. 2013	CBCT, lateral cephalograms, profile photographs	Maxilim, Dolphin	yes	25	15	4 months	Error range (horizontal plane): −1.41 to 1.20 mm (Dolphin) −1.60 to 1.50 mm (Maxilim) Error range (vertical plane): −1.85 to 1.55 mm (Dolphin) −4.25 to 2.42 mm (Maxilim)	Errors with Dolphin are higher than with Maxilim, but not statistically significant ($p >$ 0.05). Both showed inaccuracies (chin, lips) for complicate surgeries

3.2.1. Works Based on Traditional (2D) Methodologies

Since a certain number of 3D software currently available for clinical practice are based on algorithm developed to work on 2D and then adapted to 3D representation of data, a section dedicated to traditional soft-tissue prediction is introduced. Moreover, since many surgeons currently rely more on traditional techniques than on three-dimensional methods, it is reasonable to focus on the limitations of this technology; 2D analysis based on lateral cephalograms is subject to errors caused by the impossibility to accurately locate points on both hard and soft tissue relying only on 2D anatomic structures [7]. Analysis performed on cephalograms reported that reasonable prediction can be made on soft-tissue landmarks positions, but with low accuracy concerning the lips and particularly in the vertical prediction of displacements [8]. Moreover, the inaccuracy increases with a higher level of complexity of intervention, which leads to errors both in hard- and soft-tissue simulation [9].

In this section, we particularly focused on study assessing the accuracy of Dolphin Imaging, but we also reported studies that took into consideration other software.

Peterman et al. worked on the quantification of the accuracy of Dolphin VTO (visual treatment objective) in predicting soft tissue changes in patients with a class III deformity and on the validation of its efficacy. In their study, authors considered fourteen patients receiving comprehensive orthodontic treatment and orthognathic surgery including both maxillary advancement and/or mandibular setback. Cephalometric tracing and analysis were performed with Dolphin Imaging software; pre- and post-operative and traced cephalometrics were superimposed using the cranial base as reference. The maxillary movements were recorded at the anterior nasal spine (ANS) and A point, while the mandibular changes were recorded at B point and pogonion (Pg) in both x and y axis. Subsequently, the pre-treatment profile pictures were superimposed to digitally-traced soft-tissue landmarks of pre-treatment cephalometric radiograph to initiate software VTO simulation (Figure 1). Finally, a prediction profile photograph generated by the software was compared to actual post-operative patient's profile photograph; coordinates of nine soft-tissue landmarks were used to quantify on each axis the differences between the predicted and the actual of patients. The results reported in this study are consistent with the prevision error calculated in previous works; Dolphin Imaging software was found to have a much larger standard deviation in the Y direction than in the X, with various degrees of accuracy in both directions for all soft tissue landmarks. The accuracy was 79% and 61% for X-axis and Y-axis respectively, with an acceptable error set at 2.00 mm. Furthermore, for these surgical cases, the least accurate prediction was the one concerning the lower lip region. In conclusion, authors affirmed that VTO program can be useful for rough movement previsions during surgical planning, but surgeons cannot rely on it for precise surgical movements (measurement range lower than 1.00 mm) [10].

Figure 1. Super-imposition of pre- and post- and traced cephalometric tracing showing advancement of the maxilla and setback of the mandible [10].

In the study of Pektas et al., the treatment planning for each patient was constituted on the base of clinical and cephalometric evaluations, and pre-operative study models. All patients were treated with pre- and post-surgical fixed orthodontic appliances. Scanned images of cephalometric radiographs were processed using Dolphin Imaging software; the Ricketts and Steiner analysis was selected to start the digitization. To compare actual outcome and software simulation, firstly the superimposition of pre- and post-operative cephalograms was performed on the sella-nasion plane registered at sella. Landmarks located on structures not involved in the surgery were directly transposed from pre-operative to post-operative cephalogram to refine the superimposition process. The objective of an accurate superimposition of pre- and post-operative cephalograms was to construct a spreadsheet for landmarks movements to quantify changes of each point after operation; the position of points was defined in form of x and y. In accordance with the spreadsheet, a simulation of the treatment was produced, and the prediction tracings were obtained from actual post-operative changes amount. Actual post treatment tracings and prediction tracings were superimposed and differences have been measured again according to an x, y coordinate system.

The mean differences between the predicted and the actual result were less than 1 mm in four of seven soft tissue measurements; predictions turn out to be more accurate for the sagittal plane than for the vertical one. The largest errors were reported for the lower lip region, in accordance with other studies considered by authors and investigating other software [11].

Ana de Lourdes sá de Líra et al. studied the accuracy of digital prediction with respect to the actual post-operative outcome with Dolphin program. Patients with class II malocclusion were involved. Eighty Caucasian subjects treated with combined surgical and orthodontic treatment were considered. Cephalometric radiographs of pre- and post-operative were digitized; a time of at least one year from surgical to post-operative was required to rule out any effects of post-operative swelling. Digital tracings were performed with Dolphin imaging software and digital lateral cephalometric radiographs were traced at the same time to minimize the error variance; before tracing, Dolphin program was used to carry out the compensation for the effect of radiographic magnification. An x y coordinates system was constructed with the x axes corresponding to the horizontal Frankfort plane and the y-axes passing through Nasion and perpendicular to x-axes. On x-axes, vertical changes have been measured, while on y-axes authors evaluated horizontal changes. Dalberg formula was used in evaluating the reproducibility of measurements; with regard to the method error, it did not exceed 0.37° for angular measurements and 0.29 mm for linear measurements. Results of the statistical analysis were based on mean values found in each studied group. The analysis of the authors underlined that there were no significant differences between soft-tissue values at pre-operative (t1), post-operative (t2) and predicted (t3) stages. In both groups, surgeries have been more extensive than planned; facial convexity and the distance between the lips and the cranial base presented similar values between t2 and t3. The conclusion of the authors was that statistical differences in the considered measurements did not invalidate surgical prediction performed with the software, because forecasted changes were sufficiently close to of the surgical outcomes [12].

Lira et al. followed the same approach of the previous article on two groups of patients; the former underwent Le Fort I without segmental surgery, while on the latter, in addition to LFI surgery, mandibular setback was performed with an SSRO. What came out from this study was that 12 months after the operation surgical reductions in mandibular length and angle were substantially greater than indicated by predictive cephalometric tracings. The analysis of dental and skeletal mandibular changes in anteroposterior and vertical directions revealed that the effect on the profile was greater than those of the maxilla; this is probably due to the fact that the mandible is a moving structure and the upper lip leans upon the lower lip. Maxilla advancements were performed for both groups according to the planning provided by Dolphin imaging software; in group two, mandibular setback was statistically overestimated by the software planning. Nevertheless, authors concluded that statistically the differences in the considered measurements do not invalidate the surgical prediction generated with the software, because the predicted result was close to the surgical outcome. Moreover, they noticed that concerning the mandible, dentoskeletal measurements in horizontal and vertical directions showed a greater correlation to the profile than the maxillary measurements [13].

Magro-Filho et al. worked on a subjective comparison of the soft-tissue surgical simulation of two software: Dentofacial Planner Plus (Dentofacial Software, Toronto, Ontario, Canada) and Dolphin Imaging (version 9.0). Surgeries considered in the study involved retro-positioning of the mandible and advancement of the maxilla, in which linear movements were at least 4 mm for one bone segment or in the sum of both mandibular and maxillary movements, without, or least possible, postsurgical orthodontic movements. Profile predictions were made six months after the surgery; real movements, obtained from the superimposition of the cephalometries of pre- and post-operative of each subject, were used as values of the prediction. Soft-tissue images of pre-operative, real predictive images of Dentofacial Planner Plus (DFP) and Dolphin Imaging (DI) and post-operative images were compared (Figure 2). With Photoshop program (version 6.0, Adobe, San Jose, Calif), pre- and post-operative profile images were digitized and standardized; then, images were imported in DFP and DI. Movements of maxilla and mandible were made using the real values extracted from the comparison between pre- and post-operative lateral cephalograms of the surgical displacements along the vertical and horizontal directions for molars and incisors of the maxilla and for incisors of the mandible. The tip of the nose, the nasolabial angle, upper and lower lips, menton region, the base of the mandible and the complete profile were used to compare simulated and real images. A group of orthodontists, maxillofacial surgeons and general dentists evaluated the similarity among the programs simulated images and the real post-operative images with a scale ranging from "very similar" to "different". For the comparison of soft-tissue cephalometric points, to compare the two programs and to judge the criteria of similarity, chi-square test was used. From their analysis, the authors reported that for nasal tip, chin area and mandibular base DI gave better results, while for nasolabial angle, upper lip and lower lip DFP produced a more accurate prediction; authors also reported that there was no difference in the evaluation of the complete profile. Moreover, in terms of time, working with DI was longer than working with DFP. In conclusion, it was stated that in treating subjects affected by dentofacial deformities, it would be better if orthodontists and oral maxillofacial surgeons were to base on their clinical experience and to use software just as a co adjuvant. Considering the methodology of the study, there were not many differences in obtaining a two-dimensional prediction of patient's profile [14].

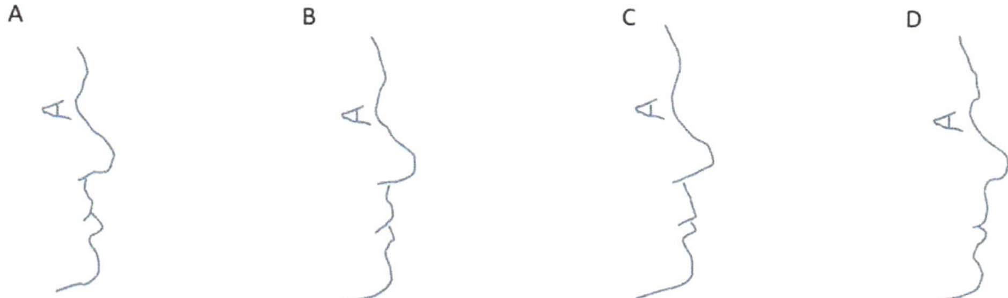

Figure 2. Profile images of a patient with Class III malocclusion, treated with orthognathic surgery: (**A**) pre-treatment; (**B**) simulation from Dentofacial Planner Plus (DFP); (**C**) simulation from Dolphin Imaging (DI); (**D**) actual post-operative photograph [14].

Other software are available for the prediction of soft tissue changes. OrthoForecast is a data-based prediction software; its database contains information of 400 patients (100 pts with asymmetry, 100 pts with skeletal class II jaw relationship and 200 pts with skeletal class III jaw relationship). All patients underwent SSRO or IVRO, with or without LFI osteotomy (respectively, 352 and 48 cases). It was evaluated by Abe et al.; the study included 15 patients with facial asymmetry, 15 with skeletal class II jaw relationship and 15 with skeletal class III jaw relationship. Patients underwent LFI osteotomy,

SSRO and IVRO. Twenty-four landmarks were chosen and digitized. Some angles were also compared between the following lines: line 1, the line connecting the bilateral mandibular angles; line 3, the line connecting the bilateral eyespots; line 2, the line connecting the bilateral oral angles. In the asymmetry group, they measured angle 1, identified by line 1 and line 3 and angle 2, identified by line 2 and line 3 on the frontal facial photograph to assess the asymmetry of the face. In the class II and class III groups, the value of angle 3, identified by the angle formed by lines connecting eyespot, upper lip, and soft tissue B-point on the lateral facial photograph was measured. Angle 3 was the reference for evaluating the convexity of the face, which was originally defined as soft-tissue convexity (Figure 3). The distances measured between landmarks on the actual and on the predicted images were analyzed, and the similarity between real and software-generated images was assessed by 39 evaluators. The mean difference measured between the positions of landmarks is less than 3.4 mm and less than 1.0° in the evaluated angles. Moreover, more than half of the evaluators stated that, in all groups, the predicted images were very 'similar' or 'similar' to the actual situation of the patients in the post-operative condition. Less than 6% of the evaluators ranked the predicted images as 'different' from the actual outcome. In conclusion, OrthoForecast is considered by the authors to be a software with high levels of accuracy, reliability and usefulness [15].

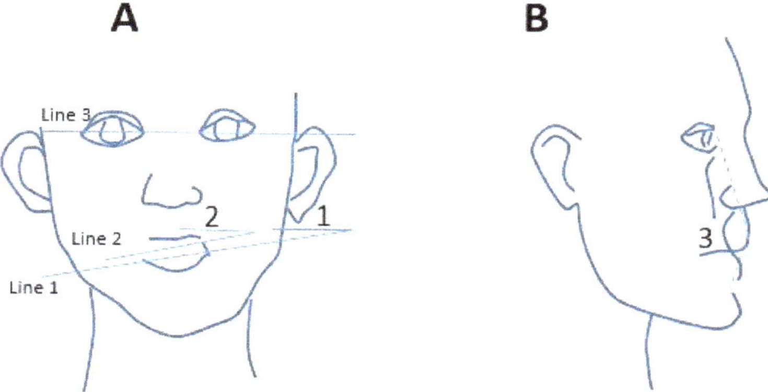

Figure 3. Construction of feature points and lines (**A**) 1, Angle between line 1 (connecting the bilateral mandibular angles) and line 3 (connecting the bilateral eyespots); 2, angle between line 2 (connecting the bilateral oral angles) and line 3; (**B**) 3, angle formed with lines connecting the eyespot point, upper lip point, and soft tissue B-point [15].

Donatsky et al. evaluated another prediction software, TIOPS. This is a computerized, cephalometric, orthognathic surgical planning system previously used in studies on hard tissue stability and accuracy. As other soft tissue prediction systems, TIOPS is based on predefined ratios of hard to soft tissue movements; its present algorithms partly rely on cephalometric observation of the post-operative situation. The study included 52 patients. Clinical photographs, study models mounted on an articulator (SAM) and standardized lateral cephalometric radiographs of the pre-operative situation were performed. Standardized lateral cephalometric radiographs were performed 5–6 weeks after surgery. The mean accuracy of the planned and predicted results of both hard and soft tissue varied from 0.0 mm to 0.5 mm from one cephalometric reference point to another. In the locus of cephalometric reference points, where it was shown that there were significant differences between planned/predicted hard and soft tissue changing in terms of position, these significant inaccuracies were moderately small and varied from 0.2 mm to 1.1 mm, with the exception of the horizontal position of the lower lip. However, the variability of the previewed hard and soft tissue individual result was evaluated to be relatively high. The study demonstrated moderately high predictability of the immediate post-surgical hard and soft tissue outcome. However, due to the significant variability of simulated individual results, it would be better to be cautious in presenting the planned and predicted hard and soft-tissue results to the individual patient in the pre-operative [16].

Finally, Chai Hui Koh et al. evaluated the accuracy of soft tissue predictions generated by the CASSOS (Computer-Assisted Simulation System for Orthognathic Surgery) software in Chinese skeletal class III patients underwent bimaxillary surgery. The digitization of pre-surgical and post-surgical lateral cephalograms of 35 patients was performed with the CASSOS program. A simulation of the surgery was performed on the pre-surgical tracing. An analysis was performed on thirty-two linear measurements on the cephalograms superimposition to assess the differences in the soft tissue profile between the post treatment results and the predicted result. It showed differences on 16 linear measurements with the most prediction errors on the upper and lower lip having a mean difference relatively small, with a greatest mean difference of 2 mm in the vertical position of stomion inferius. The authors concluded that CASSOS 2001 produced a clinically meaningful forecast of soft-tissue profile changes following bimaxillary surgery [17].

The same CASSOS software was evaluated by Jones et al. for 33 patients affected by class III skeletal deformities and underwent maxillary advancement (17 patients) or bimaxillary surgery (16 patients) [18]. The post-operative cephalograms were used to determine surgical bony movements to produce the CASSOS profile simulation. Linear differences between the predicted profile and the real outcome were measured at 12 soft tissue landmarks. The authors' conclusion was that the profile predictions obtained with CASSOS could be considered useful for both type of surgeries, even if substantial variations were found. As previously [17], the most inaccuracies were found in the lip region.

3.2.2. Works Based on Maxilim Software (3D)

In this section a collection of studies is presented considering patients' soft tissue prediction obtained with Maxilim software (Medicim—Medical Image Computing, Mechelen, Belgium). This software, based on the mass tensor model algorithm (MTM), allows surgeons to determine bony movements and to see the effect of the procedure [19]. MTM has been introduced by Cotin et al.; the geometry of the anatomical structure is discretized into a tetrahedral mesh in which the displacement vector at a generic internal point is defined as a function of vertices displacement vectors. The elastic force is written as a function of these same four vectors. The displacement for a set of mass points is set fixed and other model points will move as consequence of elastic forces due to the displacements of fixed points; finally, the new rest position of the free points is computed by integrating the Newtonian motion equation [20].

Shafi et al. [21] investigated the accuracy of Maxilim using cone-beam computed tomography (CBCT) scans in pre- and post-surgery phases (6–12 months after surgery) for 13 patients subjected to Le Fort I surgery. A 3D mesh was generated from skeletal movements (predicted model). Subsequently, the mesh generated from the post-operative of the patient and the predicted model were compared to evaluate the accuracy of the prediction. The soft tissue was divided in different areas: nose, right and left nares, right and left paranasal regions upper and lower lip and chin. The absolute distance was calculated between the meshes for each region. For each facial region, the absolute distance was calculated between the meshes.

In almost all the facial regions, the accuracy was significantly less than 3.00 mm (3.00 mm is clinically acceptable). The upper lip area was the exception; in fact, for this region, the accuracy was greater than 3.00 mm. In all cases, Maxilim produced an over prediction of the new position of the upper lip. A possible reason could be a non-linear response of the upper lip to some hard-tissue movements and, consequently, the modeling algorithm used (MTM) hardly previewed this response.

The conclusion of the authors was that the 3D soft tissue prediction for Le Fort I advancements produced by Maxilim was in general clinically satisfactory, but it was associated with marked errors around the region of the upper lip.

Liebregts et al. evaluated the accuracy of Maxilim for 3D simulation of soft tissue changes after bimaxillary osteotomy. The 3D rendered pre- and post-operative scans were matched. Segmented maxilla and mandible were aligned to the post-operative position. In order to calculate the error between the simulation and the actual post-operative condition, authors used 3D distance maps and cephalometric analysis. Concerning the facial profile, the mean absolute error between the 3D simulation and the actual post-operative facial profile was 0.81 ± 0.22 mm for the face as a whole. In this study, the accuracies of the simulation (average absolute error ≤2 mm) for the whole face and for the upper lip, lower lip and chin sub regions were 100%, 93%, 90% and 95%, respectively.

Authors affirmed that the MTM based soft tissue simulation is an accurate model for the prediction of soft tissue changes following bimaxillary surgery. The magnitude of skeletal movements influences the accuracy of the prediction; moreover, the age of the patient and the use of V–Y closure affect the precision of the predicted model. Low predictability on the upper and lower lip regions is registered once again [22].

The same authors evaluated Maxilim performances for soft-tissue simulation in 100 patients underwent BSSO for mandibular advancement [23]. As in the previous case, the accuracy of the simulation was assessed with two methods, a 3D cephalometric analysis and a 3D distance map for the entire face and for specific regions of interest. Their analysis showed that for the entire face the mean absolute 90th percentile error was less than 2 mm (clinically acceptable); the least accuracy resulted in the region of the lower lip, while the most accurate prediction involved the sub nasal region. A possible explanation suggested by authors for labial inaccuracies was the difficulty to replicate labial position during different image acquisition.

The conclusion drawn by authors on the basis of this analysis was that the soft-tissue prediction produced by Maxilim software was clinically acceptable both for the whole face and for restricted regions, but the limitation was the fact that patients considered in the study underwent a single type of surgery instead of combined operations.

In the analysis by Mundluru et al., 13 non-syndromatic adults with a midline deviation of the chin point not less than 2.0 mm underwent Le Fort I or bimaxillary osteotomies (BSSO) to correct facial asymmetry. The accuracy of an innovative concept for the soft-tissue changes prediction was evaluated. The segmentation of the 3D model was performed to identify the following regions: upper lip, lower lip, chin area, right and left paranasal regions, nose and right and left cheeks. To evaluate the prediction reliability, the surface distances between predicted and actual post-operative positions were measured. Particularly, mean (signed) and mean absolute distances were calculated on 3D meshes for each region. Through a directional analysis, the accuracy of the prediction of soft-tissue changes was evaluated. The results showed that the distances between the predicted and the actual post-operative soft tissue models were less than 2.0 mm in all regions. In details, a general tendency to the under prediction was found in the area of the cheek and of the chin (medio-laterally). An over-prediction of changes was found at the inferior border of the mandible (bilaterally) [24].

De Riu, et al. realized a retrospective study with the purpose of giving an objective quantification of accuracy of the Maxilim virtual planning method, through the comparison between planned and actual movements in jaw osteotomy. They started from the assumption that a simple superimposition of simulation and cephalometry results did not allow to define a correlation between positional error and surgical movements. Indeed, a slight positional error could be insignificant talking about large movements, but could be unacceptable dealing with small displacements. A 3D planning was performed, and virtual frontal and lateral cephalometries of pre-operative and simulated surgery were extracted; those were compared to define the predicted movements of the jaw in reference to cranial bones. All surgeries have been planned by the same surgeons using Maxilim software, with digital intermediate splints to guide osteotomies. Cephalometric analysis has been performed with Dolphin. To evaluate the accuracy, the mean linear difference between planned and actual movements was considered. Measurement of the differences between planned and actual movements were deemed accurate. As reported by authors, most of the clinical studies validating the virtual surgical planning set to 2 mm the criterion for the prediction success, with a success rate slightly lower than 100%. In this study, some significant differences with other works were found; first, differences between planned and achieved anterior facial heights were found ($p = 0.033$). This discrepancy was relevant ($p = 0.042$) in patients that did not undergo genioplasty, while it was not significant in patients who underwent this surgery (0.235). This error is probably due to the approximation of the soft tissue model, which does not allow the management of the vertical dimension. Presumably, genioplasty compensate the error on facial eight. The second error reported by authors was in the differences between actual and planned measurements for SNA (angle between Sella, Nasion and A point) and for SNB (angle between Sella, Nasion and B point), with $p = 0.008$ and $p = 0.006$ respectively. In the opinion of the authors, virtual planning inaccuracies were principally caused by the difficulty of simulate the soft tissue changes. Moreover, they concluded that virtual planning could not relieve surgeons of the necessity of monitoring jaws movements intraoperatively and of a real time supervision of planned and actual outcome, despite a high level of accuracy for most of the analyzed parameters [25].

The double blind prospective study performed by Van Hemelen et al. had the aim of providing a comparison between a traditional planning method and a 3D planning performed with Maxilim, both for hard and soft tissue [26]. In their analysis, authors considered 66 patients affected by class II or class III angle malocclusion underwent bimaxillary osteotomy (46 patients), BSSO (17) patients and LeFort I osteotomy (3 patients). Genioplasty was performed on 28 patients. For the 2D planning, clinical facial examination, lateral and frontal cephalograms were taken; 17 and 16 cephalometric landmarks were considered for 3D and for 2D respectively, to perform the validation of both 2D and 3D planning. The traditional planning was performed with Onyx Ceph Version 3.1.111 and the analyses of accuracy was performed by measuring projections on x and y axis of distances between the 16 selected points on the planned and on the actual post-operative. In the case of the 3D planning, the landmarks were defined in three dimensions and the planned and post-operative data were aligned. Distances between the 17 cephalometric points defined on planned and actual models were evaluated in depth, height and distance in the sagittal plane.

As a result of the analysis, authors reported that, for hard tissue planning, the mean differences occurring between actual and planned outcome evaluated at cephalometric points were 1.71 mm and 1.42 mm in the horizontal direction and 1.69 mm and 1.44 mm in the vertical direction, for 2D and 3D respectively. Concerning soft tissue displacements, the mean values computed in the horizontal direction were 2.29 mm for traditional planning and 1.48 for Maxilim prediction; in the vertical direction, authors noted mean values of 2.07 mm and 1.46 mm for 2D and 3D techniques respectively. These evaluations led authors to conclude that there was a statistically significant difference in the soft-tissue prediction between the two considered methods (chi-squared test and independent t-test were used to determine statistical significance, with $p < 0.005$); particularly, the 3D approach seemed to be more predictive than the traditional one. No statistically significant differences were reported for hard tissue planning.

The conclusion reached by authors was that, concerning the soft tissue prediction, the 3D planning approach here performed with Maxilim software led to a more accurate planning than the traditional method, even with disadvantages such as the cost of the software and of the CBCT scans. Moreover, a higher learning time must be considered when planning with the 3D method. In the authors' opinion, 3D planning will not replace the traditional prediction if those negative aspects are not addressed.

The study by Nadjmi et al. focused on the accuracy of soft-tissue profile changes simulated by Maxilim with respect to the accuracy obtained with Dolphin. Particularly, the goal of the comparison was to assess if the 3D prevision was more accurate than the 2D one. A "natural head distance" was defined as distance between suprasternal notch and soft tissue pogonion to replicate patients' head position. On lateral cephalogram, soft tissue and hard tissue landmarks (15 and 25 points respectively) were located; soft tissue predictions were generated with Dolphin and Maxilim. The two predictive results were superimposed with patient's post-operative profile photograph and differences in terms of linear measurements of landmarks in x and y directions were measured. All comparison steps were performed by the same operator. The movements of bony structures in the simulation must be the same as performed by the surgeon, to correctly compare the two predictions. Since this study made a comparison between 2D and 3D simulations, even if Maxilim gave in output 3D models, only profile (2D) images were considered. From the analysis of the results, the authors found that both Dolphin and Maxilim predictions were accurate, but the simulation obtained with Maxilim made possible a quantification of volumetric changes (lips aspect) and a prevision of changes in the transverse plane. Even if the study focused on the comparison of accuracies (and for this reason only the lateral prediction was considered) between the two software, authors stated that 3D simulation would be helpful in complicated surgeries thanks to the third-dimensional information [27].

3.2.3. Works Based on Other 3D Software

The work of Resnick et al. investigates the accuracy of Dolphin 3D soft tissue prediction using seven patients who had a single-segment Le Fort I osteotomy. Dolphin 3D Imaging (Dolphin Imaging and Management Solutions, Chatsworth, CA, USA) is based on a landmark photographic morphing algorithm developed for 2D prediction and adapted to 3D. The user must locate 79 landmarks, 47 on the hard tissue and 37 on the soft tissue, on the CT volume; the system generates adaptable curves between the landmarks, similarly to the tracing of a lateral cephalometry.

On the pre-operative segments LF I osteotomy was simulated by Dolphin, then hard and soft tissue landmarks were assigned, morphing curves were adjusted and finally prediction image (Tp) and post-operative image (T1) were aligned and registered for the measurement of the differences. For each patient, the error was calculated as the difference occurring between Tp and T1 at 14 points, six on the midline and eight laterally and at the nasolabial angle (Figure 4). Twelve of these points are standard anthropometric or cephalometric landmarks. Two new points were introduced for the study: lateral ala (LA), obtained from the intersection of the line tangent to endocanthion (EN) and the one tangent to subalare (SBAL), and maxillary buttress (MB), traced at the intersection of lines tangent to exocanthion (EX) and SBAL (Figure 4A).

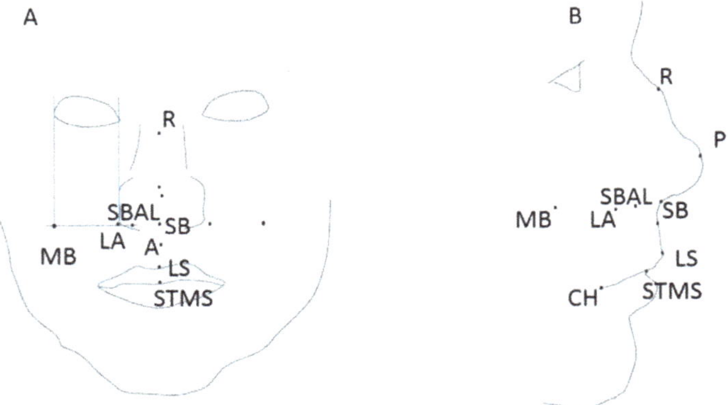

Figure 4. Points for measurement by Resnick et al. (**A**) Frontal view. The two points derived for this study are shown: lateral ala point (LA), defined as the intersection of lines tangent to endocanthion (EN) and subnasale (SBAL), and maxillary buttress (MB) point, and defined as the intersection of lines tangent to exocanthion (EX) and SBAL. (**B**) Lateral view showing all points [28].

In closing, their analysis showed that the capability to predict the soft-tissue changes in the three dimensions after LF I surgery using Dolphin software had some limitations. For linear changes, its accuracy was acceptable, while it was not acceptable for lateral points of the face. Concerning the midline, changes at nasal base were more subject to prediction errors [28].

Bianchi et al. [29] and Marchetti et al. [30] worked on SurgiCase_CMF using CBCT and MSCT respectively of 10 patients with facial deformities. In the study presented by Bianchi et al., patients underwent LFI, BSSO and genioplasty. The pre- and post-operative CT images were aligned on the top of each other. Firstly, a registration point was made, with the location of two corresponding landmarks on pre- and post-operative conditions. After the manual registration, automatic surface registration was carried out using an iterative closest point algorithm. A crucial point in surfaces registration was to indicate those that did not change after the surgery; authors used the eyes area and the forehead as fixed regions. Finally, the post-operative soft tissue surface was compared with the software virtual simulation. A comparison algorithm was used to measure the distance of every triangle corner of the post-operative surface against the preoperative one. The percentage of error was <2.0 mm for the 86.80% of patients, even if there were important errors in the top lip and chin regions. The results in the group of patients studied with CBCT (which had a reliability of 86.8%) stated that the use of SurgiCase_CMF 1.2 software combined with CBCT data allowed to achieve a realistic preview with low X-ray exposure [29].

Marchetti et al. divided virtual surgical planning in four steps (Figure 5): (1) CT data reconstruction; (2) generation of 3D models of hard and soft tissues; (3) various virtual surgical planning and simulation mode; (4) different pre-operative previews of the soft tissues. Skeletal model surgical planning and simulation were performed on the base of the 3D CT; the soft tissue model was made through a physical model. According to clinical options, the software generated a set of simulations and models of the soft tissues; in order to prevent temporomandibular functional problems, an orthodontist assessed the pre-operative plans. From the comparison of simulation and CT surgical outcomes, it was found that the prediction of soft tissue situation had a reliability greater than 91%, with a percentage of error lower than 2 mm for 76%–99%. According to the obtained results, SurgiCase_CMF was able to give a realistic and accurate preview of the face of the subjects undergoing surgery, but there were significant errors in the lip and chin regions, as in the previous studies [30].

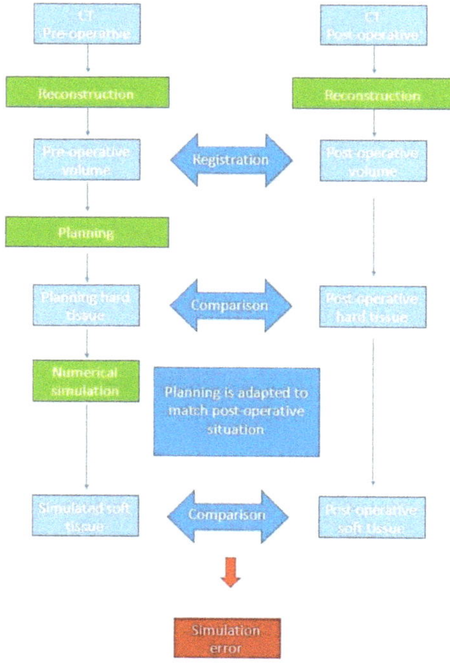

Figure 5. Virtual surgical planning proposed by Marchetti et al. [30].

3dMDvultus is a software package in which the rendering function is based on Mass Spring Model. MSM assumes a discretization of a deformable object into n mass points and a set of m connections between each n mass point. In a tetrahedral discretization, for each mesh node a point is allocated, and it is defined a linear spring for all the edges of the mesh. This linear spring follows Hook's law. This model has the advantage of being simple and computationally efficient, but the disadvantage is that the elastic behavior of the model is determined by the spring constant; the value of this constant is an approximation and it has no true bio-mechanical relevance [19].

Khambay et al. used 3dMDvultus to predict the soft tissue changes after surgery. The work included 10 patients who underwent Le Fort I osteotomy. Ten landmarks were used to measure the distances between the pre- and post- operative meshes. Moreover, the percentage of mesh points minor or equal to 2.0 mm were calculated for the full face and for specific anatomical regions. The results demonstrated that the percentage of mesh with errors minor or equal to 2.0 mm for the full face was 85.2%–94.5% and 31.3%–100% for anatomical regions. The range of root mean square (RMS) error was from 2.49 mm to 0.94 mm. The most of mean linear distances computed between the surfaces was equal or less than 0.8 mm, but it increased for the mean absolute distances. From this analysis, it was stated that the use of specific anatomical regions was more significant in clinical practice than the use of full face [31].

As in the previous article, the software considered by Terzic et al. is 3dMDvultus. Three-dimensional photographs of patients' head in natural position (lips and muscles at rest, open eyes, neutral facial expression) of pre- and post-operative were taken; CT and/or CBCT images were imported to the software platform. Pre-operative 3D photograph was fused with pre-operative CT/CBCT images. In the same manner, post-operative bony skull data was fused with pre-operatives by matching bone areas not involved in surgeries. According to actual post-surgical CT, 3D bone segments were reproduced in the pre-operative skull; the software rendering function (mass spring model) was activated. Osteotomy segments were moved to real post-operative position and the soft-tissue rendering generated the textured facial soft tissue prediction. After the fusion of 3D photographs of predicted and real post-operative soft tissue, a horizontal plane was positioned arbitrarily to divide the face in an upper part not involved in surgery, and a lower part including the overlap of simulated and real post-surgery result. Averaged distribution of absolute error showed more discrepancies between predicted and real post-operative outcome in the lower part of the face: in 29.8% of lower halves authors found errors exceeding 3.00 mm. This preliminary study concluded that the software platform had an insufficient accuracy in the forecast of 3D soft-tissue displacements [32].

Ullah et al. carried out a retrospective study in which they focused on the ability of 3dMDvultus to give a prediction of the face appearance of patients undergoing LFI maxillary advancement. The null hypothesis they started from was that the mean difference in absolute distance between the predicted surface generated by the software and the real 3D facial surface measured at eight anatomical regions did not exceed 3 mm. Pre- and post-operative CBCT for each patients were imported into 3dMDvultus and hard tissue and soft tissue were separately segmented and saved in STL (standard triangulation language) format. A template of actual surgical changes was produced using a CAD/CAM software (VRMesh, VirtualGrid, Seattle, WA, USA). Post-operative hard and soft tissue linked together were aligned to the anterior cranial base of pre-operative hard tissue of the same patient. Regions of maxilla and mandible of post-operative models were saved as STL file and imported into 3dMDvultus to generate a template of the actual maxillo-mandibular complex aligned in the same 3D space as the pre-operative image. Using the same software, a LFI osteotomy was simulated for the pre-operative hard tissue. The resulting soft-tissue prediction was exported as an STL file for the analysis. The predicted model and the real surgical result were imported as mesh in VRMesh; the differences between the two models were rendered as a color map, which showed the overall accuracy of the software. After the analysis on the overall surface, the accuracy was evaluated subdividing the mesh into anatomical regions: chin, lower lip, upper lip, nose, right and left nares, right and left paranasal areas. From the analysis of the results, the authors concluded that since the distances between software-generated surfaces and real facial surfaces were lower than 3 mm, 3dMDvultus 3D soft-tissue predictions were acceptable for clinical use in LFI osteotomy. MSM seemed to correctly predict the positions of lip and chin, while improvements should be made for nasal and paranasal regions [33].

In the prospective study by Holzinger et al., 16 patients with open bite dentofacial-dysmorphosis and underwent orthognathic surgery (surgery first) were analyzed. The surgery was planned using conventional sketches as a new software developed by the authors, SOTIRIOS planning software. A conventional pre-surgical planning was carried out using three sides X-rays and, in addition, a CT scan was performed before and 6 months after surgery. A constrained fitting approach using statistical shape modeling (SSM) technique to generate the patient-specific data was used in combination with the anatomical landmarks to compute the patient specific model. A quantitative comparison of the soft tissue prediction and post-operative data was made showing a mean error of 1.46 mm ± 1.53 mm. Authors concluded that the SOTIRIOS planning software was quite accurate and enabled the surgeon to predict the soft tissue outcome [34].

A comparison between Dolphin 3D, ProPlan CMF and an in-house probabilistic finite element (PFEM) [6] was carried out by Knoops et al. [35]. The retrospective study included seven patients who underwent LFI maxillary advancement. From CBCT taken preoperatively, 3D models of bone and soft tissue were constructed. On the base of the advancement and rotation movements measured on the post-operative images, a virtual Le Fort I osteotomy was performed and three different soft tissue predictions were realized with the three softwares. Firstly, the three predictions were compared with the pre-operative model to evaluate differences. After that, the three predictions were compared with the real post-operative outcome, to assess which was the most predictive model. As the patients involved underwent a LFI maxillary advancement, the areas of interest for the comparison with post-operative situation were the upper lip and the paranasal regions. From the outcomes analysis, it resulted that the prediction generated by Dolphin 3D was affected by a general under-prediction of paranasal displacements, while both ProPlan CMF and the PFEM model were characterized by an over-prediction of displacements involving cheilion regions. Average root mean square distances and average percentage of points less than 2 mm between real post-operative images and predictions were computed (Table 3); statistically significant differences resulted from the Friedman test. In particular, the post hoc Wilcoxon signed-rank test proved that RMS for PFEM and RMS for ProPlan had statistically significant lower values with respect to RMS for Dolphin 3D ($p = 0.016$). On the opposite, the differences between ProPlan and PFEM could not be considered statistically significant ($p = 0.219$).

The conclusions of the authors was that ProPlan CMF and PFEM gave soft-tissue predictions with significant accuracies, particularly when using the correct post-operative maxillary position; because of the limitations introduced by a landmark-based algorithm and a sparse architecture, Dolphin 3D resulted to be inaccurate when surgery involved large maxillary advancements, particularly on lateral points.

Nam et al. [36] tried to assess if a 3D virtual surgery could accurately predict the soft tissue outcome of 29 patients who underwent bimaxillary orthognatic surgery (LFI, BSSO and genioplasty) performed by the same surgeon. The predicted outcome was generated using the Simplant Pro program. All simulations were performed by the same operator. The produced simulation of the soft tissue result was superimposed on the post-operative of the patients, and 10 landmarks were designed and positioned on the two models; the simulation error was determined by measuring distances (x, y, z) between the same landmarks on predicted and real images. From this analysis, an accuracy of 52.8% resulted, disagreeing with previous study reporting accuracies around 80%. Most of the errors were found for landmarks of lower lips, mouth corner and chin area (pogonion and menton), with a general larger error in points located on the mandible. Authors suggested as a possible cause of this trend the complexity of the surgery, involving more movements in the mandible rather than in the maxilla; it could determine lower errors in the maxilla but higher inaccuracy in the mandible. Inaccuracies reported for 18 of the 30 soft-tissue landmarks were considered statistically significant ($p < 0.05$), with errors concentrated at mouth corners. Possible explanations addressed by the authors were the ethnic differences between the data used for developing the software (Caucasian population) and the data used in this study (Asian population); moreover, it was not negligible that errors up to 2 mm could be due to the positioning of soft tissue landmarks. In conclusion, authors stated that 3D software in soft tissue prediction could have great potential, but the accuracy must be improved.

Even if it is out of the scope of the present work, it is important to state that not only the surgical planning plays a crucial role in the optimal process of care of the patients. It is not negligible that maxillofacial surgery, as well as other branches of surgeries, could lead to medical complications such as venous thromboembolism, which can have fatal consequences when moving to the lung (DVT/PE) [37]. Moreover, surgeons must manage patients with rare disorders of hemostasis, which need particular care in the perioperative phase. The works by Simurda et al. [38,39], and Ghadimi et al. [40] have been referred to in order to deepen understanding of this topic.

4. Discussion and Conclusions

In this systematic review we have summarized twenty-four studies focused on the evaluation of soft-tissue prediction software for maxillofacial surgery. The treated software are commonly used in clinical procedures to have a preview of the patients' outcome after corrective surgery; the most common methodologies on which these software are based are the mass spring model, the finite element model and the mass tensor model, each of them with its advantages and disadvantages. MSM presents an easy architecture and a low memory usage but it has no real biomechanical foundations; FEM is more relevant from a biomechanical point of view, but it has a high computational cost and a rather long simulation time; MTM tries to combine the advantages of MSM and of FEM [18]. In particular, the lip region is the most difficult to predict, since lips do not rely on bony structures; moreover, it is difficult to replicate the same lip position from one acquisition to another. Due to this particularity, lips are supposed to be particularly involved in involuntary displacements caused by surgeries performed on different facial areas. The difficulty in accurately predicting the appearance of this sub region is a crucial point in the development of a prediction method, since it is particularly related to aesthetic self-perception and satisfaction of the patient. With respect to the traditional algorithms based on hard- to soft-tissue ratio, methodologies involving biomechanical models of viscoelasticity of the soft tissue to mimic its elastic deformation seems to be able to give a more realistic simulation. Despite the number of computer programs dedicated to three-dimensional prediction and surgical planning, maxillofacial practitioners often prefer to rely on conventional two-dimensional methods. This fact could be explained with the overall results given by these methodologies, which seem to be only partially relevant and/or reliable in the practice of daily clinical activity. Concerning the works based on Maxilim, Shafi et al. found that this software makes a clinically satisfactory prediction of soft tissue changes, but with an over-prediction of the position of the upper lip [21]; according to Liebregts et al., its accuracy is influenced by the magnitude of maxillary and mandibular movements, by the age of the patient and by the usage of V-Y closure, and these authors further reported a low predictability of lower and upper lips [22]. Moreover, a general tendency was reported towards an under-prediction in the area around the cheek and chin and towards an over-prediction at the mandibular inferior border [24]. It was suggested that Maxilim inaccuracy in the prediction of the lip region might be due to the modelling algorithm, i.e., mass tensor, and might indicate a non-linear response of the upper lip [21]. The study by De Riu et al., although not directly focused on soft tissue prediction, reports that inaccuracies in virtual planning are primarily due to the soft-tissue virtual model [25].

With regard to Dolphin software, Resnick et al. reported that the ability to predict 3D soft tissue changes was limited, with an accuracy that may be acceptable for linear changes but not for lateral facial points [28]. Peterman et al. reported various degrees of accuracy at each soft-tissue landmark in the horizontal and vertical axes; lower lip prediction was the least accurate. A similar result of low prediction accuracy for the lower lip region is reported by Pektas et al. [11]. Conversely, two articles affirmed that the statistical differences between Dolphin soft tissue prediction and the real outcome of the patient did not invalidate the software prediction, as it was close to real surgical results [12,13].

Dolphin 3D imaging uses a landmark-based photographic morphing algorithm that was developed for two-dimensional prediction and then adapted to three-dimensional prediction; this could be an explanation for its imprecisions. Moreover, it requires plotting 79 landmarks on the CT volume (42 for the hard tissue and 37 for the soft tissue); the localization of such a large number of points seems to require too much time for a stable usage in clinical daily practice.

Among the other considered software, SurgiCase_CMF was analyzed by Bianchi et al. [29] and Marchetti et al. [30]; they reported that the percentage of error was less than 2 mm for the majority of patients, but in both studies an important error in the areas of lips and chin was reported. Khambay et al., Terzic et al. and Ullah et al. used 3dMDvultus to predict soft tissue changes; in general, they reported that its use on specific anatomical regions was more meaningful than on the full face [31–33]. Moreover, they reported an insufficient prediction accuracy particularly for the lower part of the face with errors exceeding 3 mm [32], and for nasal and paranasal regions [33]. Surprisingly, prediction results seemed to be reliable for lips and chin.

The results reported for TIOPS showed significant inaccuracies in the predicted horizontal position of the lower lips and a relatively high variability of the predicted individual surgical outcome. The authors suggest care is needed when presenting the predicted image to patients before the operation [16,41]. The SOTIRIOS planning software and the CASSOS program were found to be quite accurate [34] and able to give a clinically useful prediction of the soft-tissue profile change [17]. OrthoForecast software resulted as being in general an accurate tool for soft-tissue prediction; indeed, more than half of the involved evaluators assessed the predicted images to be very similar or similar to the actual post-operative images [15].

As a result of this review, it seems that Dolphin, CASSOS and OrthoForecast are the most accurate software in soft-tissue prediction. However, their reliability is not sufficient when further surgery is carried out and the complexity of the operation increases; moreover, they provide only partial information on the prediction, due to the lack of the third dimension. From this viewpoint, all articles agree on the potentiality of 3D methodologies. Maxilim produced clinically meaningful results, but its inaccuracy in specific facial areas (particularly in the lip region) led to the conclusion that further steps are needed to give surgeons a tool for predicting facial movements (including involuntary movements) that significantly supports the planning of the intervention in terms of aesthetic success. The comparison between 2D and 3D methodologies shows that the first reach better results in terms of accuracy and surgeon satisfaction. Consequently, the latter are still not able to meet medical needs.

The analysis of these studies has led us to consider some issues that are still pending. Almost all of them report problems of reliability of the software or inaccuracy in specific areas; this shows that an ameliorative method would be necessary to improve prediction results. For this reason, it would be useful to clearly understand the way the existing software compute the tissue displacements. Moreover, it is important to notice that some of the considered software (Dolphin 3D) are not based purely on 3D methods, but they are an adaptation of a 2D algorithm to 3D data; the transfer of 2D data to a 3D representation may cause errors. This aspect affects both hard- and soft-tissue prediction, and it is more evident in soft tissue simulations, due to the complexity of its behavior (involuntary displacements). The tendency of surgeons to prefer conventional methods suggests that, despite the general positive opinion emerging from the studies, the reliability of automatic systems is not sufficient, since it has not significantly increased above the two-dimensional methods commonly used in clinical routine. Indeed, it seems to lack consistency between the results obtained with prediction software and practitioner satisfaction. As previously reported, if the statistical differences in the evaluated measurements do not invalidate surgical predictability software, which parameters can be used to evaluate the precision of the prediction? It may be that the evaluation of automatic output is qualitative more than quantitative. Surgeon opinion on prediction software is that, even if the soft-tissue simulation outputs are plausible with a possible result of the interventions, these do not correspond to the effective post-operative outcome; moreover, this discrepancy is even more evident when the complexity of surgery increases. Neither hard- to soft-tissue ratio-based algorithms nor those based on viscoelastic models seem to meet surgeon demand. Another aspect is that currently soft tissue prediction seems to be focused on specific intervention effects more than on involuntary displacements; presently, these displacements affect, in an unpredictable manner, the facial outcome of patients. An accurate and reliable prevision method should consider these displacements to give surgeons an effective and helpful tool.

To improve prediction, a clear technical scientific evaluation approach should be established, which seems to be lacking in certain studies. Moreover, it would be better to clarify the level of precision of the predicted outcome needed by surgeons. These aspects are also an incentive for researchers to investigate the correlation between the extent of soft tissue changes with respect to STTs (soft-tissue thickness) and BMI (body mass index).

Author Contributions: Conceptualization, F.M. and E.Z.; methodology, J.P.A.S., S.N., F.M.; software, E.C.O. and S.N.; validation, S.N., E.C.O., and J.P.A.S.; formal analysis, F.M.; investigation, S.N.; resources, J.P.A.S. and E.Z.; data curation, E.C.O. and S.N.; writing—original draft preparation, S.N., E.C.O., and J.P.A.S.; writing—review and editing, F.M. and E.Z.; visualization, F.M.; supervision, F.M., E.Z., E.V., G.R.; project administration, E.V. and G.R.; funding acquisition, E.V. and G.R.

Funding: This research received no external funding.

Conflicts of Interest: The authors declare no conflict of interest.

References

1. Bruce, V. Applied Research in face Processing. In *The Oxford Handbook of Face Perception*; Oxford University Press: New York, NY, USA, 2011; pp. 131–146.
2. Eckardt, C.; Cunningham, S. How predictable is orthognathic surgery? *Eur. J. Orthod.* **2004**, *26*, 303–309. [CrossRef] [PubMed]
3. Jo, L.-J.; Weng, J.-L.; Ho, C.-T.; Lin, H.-H. Three-dimensional region-based study on the relationship between soft and hard tissue changes after orthognatic surgery in patients with prognathism. *PLoS ONE* **2018**, *13*, e0200589.
4. Suh, H.-Y.; Lee, H.-J.; Lee, Y.-S.; Eo, S.-H.; Donatelli, R.; Lee, S.-J. Predicting soft tissue changes after orthognatic surgery: The sparse partial least squares method. *Ang. Orthod.* **2019**, *89*, 910–916. [CrossRef] [PubMed]
5. Kim, D.; Chun-Yu Ho, D.; Mai, H.; Zhang, X.; Shen, S.G.F.; Shen, S.; Yuan, P.; Liu, S.; Zhang, G.; Zhou, X.; et al. A clinically validated simulation method for facial soft tissue change prediction following double jaw orthognatic surgery. *Med. Phys.* **2017**, *44*, 4252–4261. [CrossRef]
6. Knoops, P.; Borghi, A.; Ruggiero, F.; Badiali, G.; Bianchi, A.; Marchetti, C.; Rodriguez-Flores, N.; Breakey, R.; Jeelani, O.; Dunaway, D.; et al. A novel soft tissue prediction methodology for orthognatic surgery based on probabilistic finite alament modelling. *PLoS ONE* **2018**, *13*, e0197209. [CrossRef]
7. Swennen, G.; Schutyser, F.; Barth, E.; de Groeve, P.; de Mey, A. A new method of 3D cephalometry part I: The anatomic cartesian 3D reference system. *J. Craniofac. Surg.* **2006**, *17*, 314–325. [CrossRef]
8. Van Twisk, P.-H.; Tenhagen, M.; Gul, A.W.E.; Kuonstaal, M. How accurate is the soft tissue prediction of Dolphin Imaging for orthognatic surgery? *Int. Orthod.* **2019**, *17*, 488–496. [CrossRef]
9. Rustemeyer, J.; Groddeck, A.; Zwerger, S.; Bremerich, A. The accuracy of two-dimensional planning for routine orthognatic surgery. *Br. J. Oral Maxillofac. Surg.* **2010**, *48*, 271–275. [CrossRef]
10. Peterman, R.; Jiang, S.; Johe, R.; Mukherjee, P. Accuracy of Dolphin visual treatment objective (VTO) patients treated with maxillary advancement and mandibular set back. *Prog. Orthod.* **2016**, *17*, 17–19. [CrossRef]
11. Pektas, Z.; Kircelli, B.; Cilasun, U.; Uckan, S. The accuracy of computer assisted surgical planning in soft tissue prediction following orthognathic surgery. *Int. J. Med. Robot. Comput. Assist. Surg.* **2007**, *3*, 64–71. [CrossRef]
12. De Lira, A.D.L.S.; de Moura, W.L.; Artese, F.; Bittencourt, M.A.V.; Nojima, L.I. Surgical prediction of skeletal and soft tissue changes in treatment of Class II. *J. Cranio-Maxillo-Fac. Surg.* **2013**, *41*, 198–203. [CrossRef] [PubMed]
13. De Lira, A.D.L.S.; de Moura, W.L.; de Barros Vieira, J.M.; Nojima, M.G.; Nojima, L.I. Surgical prediction of skeletal and soft tissue changes in Class III treatment. *J. Oral Maxillofac. Surg.* **2012**, *70*, 290–297. [CrossRef] [PubMed]

14. Magro-Filho, O.; Magro-Hernica, N.; Queiroz, T.P.; Aranega, A.M.; Garcia, I.R., Jr. Comparative study of 2 software programs for predicting profile changes in Class III patients having double-jaw orthognathic surgery. *Am. J. Orthod. Dentofac. Orthop.* **2010**, *137*, 452.e1–452.e5. [CrossRef] [PubMed]

15. Abe, N.; Kuroda, S.; Furutani, M.; Tanaka, E. Data-based prediction of soft tissue changes after orthognathic surgery: Clinical assessment of new simulation software. *Int. J. Oral Maxillofac. Surg.* **2015**, *3*, 90–96. [CrossRef] [PubMed]

16. Donatsky, O.; Hillerup, S.; Bjorn-Jorgensen, J.; Jacobsen, P. Computerized cephalometric orthognathic surgical simulation, prediction and postoperative evaluation of precision. *Int. J. Oral Maxillofac. Surg.* **1992**, *21*, 199–203. [CrossRef]

17. Koh, C.H.; Chew, M.T. Predictability of soft tissue profile changes following bimaxillary surgery in skeletal class III chinese patients. *J. Oral Maxillofac. Surg.* **2004**, *62*, 1505–1509. [CrossRef] [PubMed]

18. Jones, R.; Khambay, B.; McHugh, S.; Ayoub, A. The validity of a computer-assisted simulation system for orthognatic surgery (CASSOS) for planning the surgical correction of class III skeletal deformities: Single jaw vs. bimaxillary surgery. *Oral Maxillofac. Surg.* **2007**, *36*, 900–908. [CrossRef]

19. Mollemans, W.; Schutyser, F.; Nadjmi, N.; Maes, F.; Suetens, P. Predicting soft tissue deformations for a maxillofacial surgery planning system: From computational strategies to a complete clinical validation. *Med. Image Anal.* **2007**, *11*, 282–301. [CrossRef]

20. Cotin, S.; Dekingette, H.; Ayache, N. A hybrid elastic model allowing real-time cutting, deformations and force-feedback for surgery training and simulation. *Vis. Comput.* **2000**, *16*, 437–452. [CrossRef]

21. Shafi, M.; Ayoub, A.; Ju, X.; Khambay, B. The accuracy of three-dimensional prediction planning for the surgical correction of facial deformities using Maxilim. *Int. J. Oral Maxillofac. Surg.* **2013**, *42*, 801–806. [CrossRef]

22. Liebregts, J.; Tong, X.; Timmermans, M.; de Konig, M.; Bergè, S.; Hoppenreijs, T.; Maal, T. Accuracy of three-dimensional soft tissue simulation in bimaxillary osteotomies. *J. Cranio-Maxillo-Fac. Surg.* **2015**, *43*, 329–335. [CrossRef] [PubMed]

23. Liebregts, J.; Timmermans, M.; de koning, M.; Bergè, S.; Maal, T. Three-dimensional facial simulation in bilateral sagittal split osteotomy: A validation study of 100 patients. *J. Oral Maxillo Fac. Surg.* **2015**, *73*, 961–970. [CrossRef] [PubMed]

24. Mundluru, T.; Almukhtar, A.; Ju, X.; Ayoub, A. The accuracy of three-dimensional prediction of soft tissue changes following the surgical correction of facial asymmetry: An innovative concept. *Int. J. Oral Maxillofac. Surg.* **2017**, *46*, 1517–1524. [CrossRef] [PubMed]

25. De Riu, G.; Virdis, P.; Meloni, S.; Lumbau, A.; Vaira, L. Accuracy of computer-assisted orthognathic surgery. *J. Cranio-Maxillofac. Surg.* **2018**, *46*, 293–298. [CrossRef]

26. Van Hemelen, G.; van Genechten, M.; Renier, L.; Desmedt, M.; Verbruggen, E.; Nadjimi, N. Three-dimensional virtual planning in orthognatic surgery enhances the accuracy of soft tissue prediction. *J. Cranio-Maxillo-Fac. Surg.* **2015**, *43*, 918–925. [CrossRef]

27. Nadjmi, N.; Tehranchi, A.; Azami, N.; Saedi, B.; Mollemans, W. Comparison of soft tissue profiles in Le Fort I osteotomy patients with Dolphin and Maxilim softwares. *Am. J. Orthod. Dentofac. Orthoprothes.* **2013**, *144*, 654–662. [CrossRef]

28. Resnick, C.; Dang, R.; Glick, S.; Padwa, B. Accuracy of three-dimensional soft tissue prediction for Le Fort I osteotomy using Dolphin 3D software: A pilot study. *Int. J. Oral Maxillofac. Surg.* **2017**, *46*, 289–295. [CrossRef]

29. Bianchi, A.; Muyldermans, L.; di Martino, M.; Lancellotti, L.; Amadori, S.; Sarti, A.; Marchetti, C. Facial Soft Tissue Esthetic Predictions: Validation in Craniomaxillofacial Surgery with Cone Beam Computed Tomography Data. *J. Oral Maxillofac. Surg.* **2010**, *68*, 1471–1479. [CrossRef]

30. Marchetti, C.; Bianchi, A.; di Martino, M.; Lancellotti, L.; Sarti, A. Validation of new soft tissue software in orthognathic surgery planning. *Int. J. Oral Maxillofac. Surg.* **2011**, *40*, 26–32. [CrossRef]

31. Khambay, B.; Ullah, R. Current methods of assessing the accuracy of three-dimensional soft tissue facial prediction: Technical and clinical considerations. *Oral Maxillofac. Surg.* **2015**, *44*, 132–138. [CrossRef]

32. Terzic, A.; Combescure, C.; Scolozzi, P. Accuracy of computational soft tissue predictions in orthognathic surgery from three-dimensional photographs 6 months after completion of surgery: A preliminary study of 13 patients. *Int. Soc. Aesthet. Plast. Surg.* **2013**, *38*, 184–191. [CrossRef] [PubMed]

33. Ullah, R.; Turner, P.; Khambay, P. Accuracy of three-dimensional soft tissue predictions in orthognathic surgery after Le Fort I advancement osteotomies. *Br. J. Oral Maxillofac. Surg.* **2014**, *53*, 153–157. [CrossRef] [PubMed]
34. Holzinger, D.; Juergens, P.; Shahim, K.; Reyes, M.; Schicho, K.; Millesi, G.; Perisanidis, C.; Zeilhofer, H.; Seemann, R. Accuracy of soft tissue prediction in surgery-first treatment concept in orthognathic surgery: A prospective study. *J. Cranio-Maxillofac. Surg.* **2018**, *46*, 1455–1460. [CrossRef] [PubMed]
35. Knoops, P.; Borghi, A.; Breakey, R.; Ong, J.; Jeelani, N.; Bruun, R.; Schievano, S.; Dunaway, D.J.; Padwa, B.L. Three-dimensional soft tissue prediction in orthognatic surgery: A clinical comparison of Dolphin, ProPlan CMF and probabilistic finite element modelling. *Int. J. Oral Maxillofac. Surg.* **2019**, *48*, 511–518. [CrossRef]
36. Nam, K.-U.; Hong, J. Is three-dimensional soft tissue prediction by software accurate? *J. Craniofac. Surg.* **2015**, *26*, e729–e733. [CrossRef]
37. Brice, W.; Indresano, A.; O'Rian, F. Venous tromboembolism in oral maxillofacial surgery: A literature review. *J. Oral Maxillo Fac. Surg.* **2011**, *69*, 840–844.
38. Simurda, T.; Kubisz, P.; Necas, L.; Stasko, J. Perioperative coagulation management in in a patient with congenital afibrinogemia during revision total hip arthroplasty. *Semin. Thromb. Emost.* **2016**, *42*, 689–692.
39. Simurda, T.; Stanciakova, L.; Stasko, J.; Dobrotova, M.; Kubisz, P. Yes or no for secondary prophilaxis in afibrinogenemia? *Blood Coagul. Fibrinolysis* **2015**, *26*, 978–980.
40. Ghadimi, K.; Levy, J.; Welsby, I. Perioperative management of the bleeding patient. *Br. J. Anaesth.* **2016**, *117*, iii18–iii30. [CrossRef]
41. Donatsky, O.; Bjorn-Jorgensen, J.; Hermund, U.; Nielsen, H.; Holmqvist-Larsen, M.; Nerder, P. Accuracy of combined maxillary and mandibular repositioning and of soft tissue prediction in relation to maxillary antero-superior repositioning combined with mandibular set-back. *J. Cranio-Maxillofac. Surg.* **2009**, *37*, 279–284. [CrossRef]

Article

The Effect of Patellar Tendon Release on the Characteristics of Patellofemoral Joint Squat Movement: A Simulation Analysis

Jianping Wang, Yongqiang Yang, Dong Guo, Shihua Wang, Long Fu and Yu Li *

School of Mechanical and Power Engineering, Henan Polytechnic University, Jiaozuo 454000, China;
wjp@hpu.edu.cn (J.W.); y17864208259@163.com (Y.Y.); 15538949519@163.com (D.G.);
211805020010@home.hpu.edu.cn (S.W.); fl1@gaokowl.com (L.F.)
* Correspondence: liyu@hpu.edu.cn

Received: 7 August 2019; Accepted: 11 October 2019; Published: 14 October 2019

Abstract: Objectives: This paper studies the patellar tendon release's effect on the movement characteristics of the artificial patellofemoral joint squat to provide reference data for knee joint surgery. Methods: Firstly, the dynamic finite element model of the human knee joint under squatting was established. Secondly, in the above no-release models, the release of 30% of the attachment area at the upper end, the lower end, or both ends of the patellar tendon were conducted, respectively. Then the simulations of all above four models were conducted. Finally, the results of the simulation were compared and analyzed. Results: The simulation results show that, after releasing the patellar tendon (compared with the no-release simulation's results), the relative flexion, medial-lateral rotation, medial-lateral tilt, and superior-inferior shift of the patella relative to the femur increased; the medial-lateral shift and anterior-posterior shift of the patella relative to the femur decreased. Conclusion: In this paper, the maximum flexion angle of the patella increased after the patellar tendon being released (compared with the no-release model), which indicated that the mobility of knee joint was improved after the patellar tendon release. The simulation data in this paper can provide technical reference for total knee arthroplasty.

Keywords: total knee arthroplasty; patellar tendon; patellofemoral joint; squat movement; dynamic finite element analysis

---

1. Introduction

With the application of prostheses in total knee arthroplasty (TKA), post-operative complications concomitantly usually occur [1]. One of these complications, often being ignored, is patella baja [2–4]. At the same time, relevant studies [5–9] show that, after TKA, the patellar tendon suffers certain reductions. Other studies [6,10] indicate that the patellar tendon can be shortened, the position of the patella lowered, reducing the knee joint movement and variation. A slightly low patella does not, however, affect the function of the knee joint. Check points during surgery must include patella contracture syndrome [11] and true patella baja [10], which can be treated by release of scar operation, patellar tendon lengthening, and patellar ligament insertion. Patients with pseudo-patella baja and slight joint line up, and those that do not affect knee function, are treated with follow-up observation. However, excessive upward movement of the joint line can result in serious symptoms [12], such as the need for surgery to replace a thin pad [13] and/or patellar ligament release. This paper is for studying the patellar tendon release's effect on the movement characteristics of the artificial patellofemoral joint squat to provide reference data for knee joint surgery.

2. Materials and Methods

2.1. The Establishment of a Finite Element Model

In this paper, a healthy male volunteer with a height of 173 cm and a weight of 60 kg was scanned by medical instrument computed tomography (CT) in the range of 10 cm above and below the knee joint center. The correlative parameter was set to 120 kVp and 150 mA, with a scanning interval of 1 mm. The volunteer was scanned with magnetic resonance imaging (MRI). MR (GE Sigma HD1.5T, Boston, MA, USA) was used for proton density weighting (PDWI) and chemical shift fat suppression weighted scanning, and the image coordinates of each layer of different weighted phases were consistent. Flux density is 1.5 T, and slide thickness is 1 mm [14]. Based on the CT and MRI images of the healthy volunteer, a three-dimensional geometric anatomical model including all bone tissues and main soft tissues was established (as shown in Figure 1). A three-dimensional solid model of press fit condylar (PFC) femoral prosthesis (PFC sigma, DePuy orthopaedics, Warsaw, IN, USA.) commonly used in TKA was established. Then the total knee replacement operation was simulated, and the prosthesis was assembled on the three-dimensional geometric anatomical model of the knee joint (as shown in Figure 2a). The model of the knee joint was divided into finite element meshes after TKA. In the finite element model of the knee joint, most of the unit types were hexahedrons (c3d8r, an eight-node linear brick element with reduced integration), and very few were pentahedrons (C3D6, a six-node linear triangular prism element) and tetrahedrons (C3D4, a four-node linear tetrahedron element), with a total unit number of 31,767, and 37,031 nodes (as shown in Figure 2b). The volunteer was provided details of the study and signed an informed consent form.

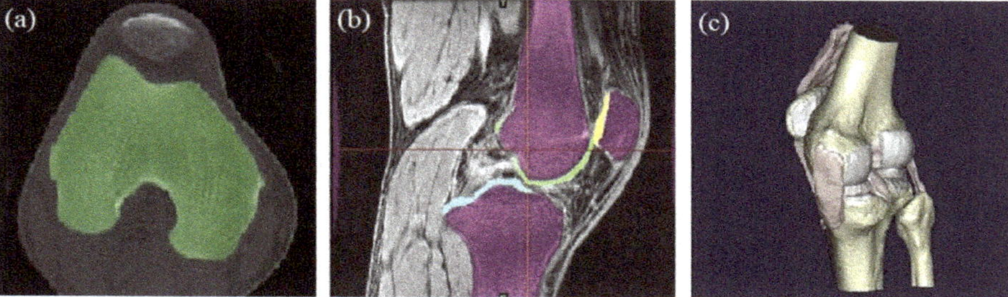

Figure 1. (**a**) Contour tracing of bone from CT; (**b**) MRI image; (**c**) the geometric anatomy model of human knee.

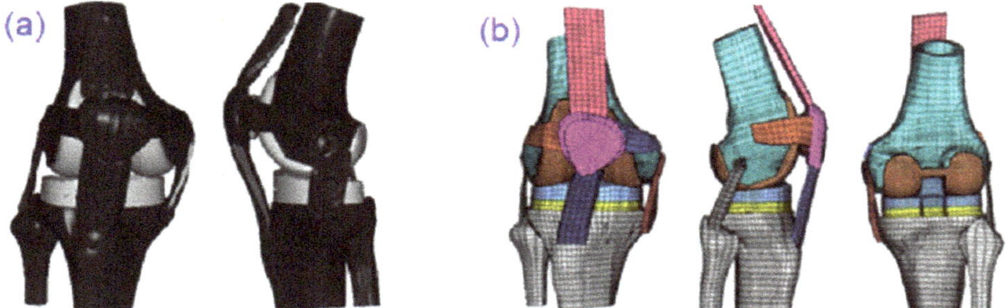

Figure 2. (**a**) 3D geometric anatomical model of knee joint after TKA; (**b**) finite element model of knee joint after TKA.

Hypermesh software, version 11.0 (Altair Engineering Corp, MI, USA) was used to simulate the release at ends of the patellar tendon. The width of the patellar tendon at the upper end was

6.84 mm before release (Figure 3a) and 4.788 mm after release (Figure 3b). The height of the patellar tendon at the lower end was 8.09 mm before release (Figure 3c) and 5.633 mm after release (Figure 3d). Figure 3 shows the finite element models before and after release the 30% of the patellar tendon at both ends (clinically, one-third of the central patellar tendon is often transplanted to repair damaged ligaments [15–18]. Therefore, 30% of the patellar tendon was used as an example). Figure 3a represents the model before release of the patellar tendon at the upper end. Figure 3b represents the model released the patellar tendon at the upper end. Figure 3c represents the model before releasing the patellar tendon at the lower end. Figure 3d represents the model releasing the patellar tendon at the lower end. The establishment of the model was based on the model built by Wang Jian-ping, and the validity of the model was verified by in vitro cadaveric experiments [19].

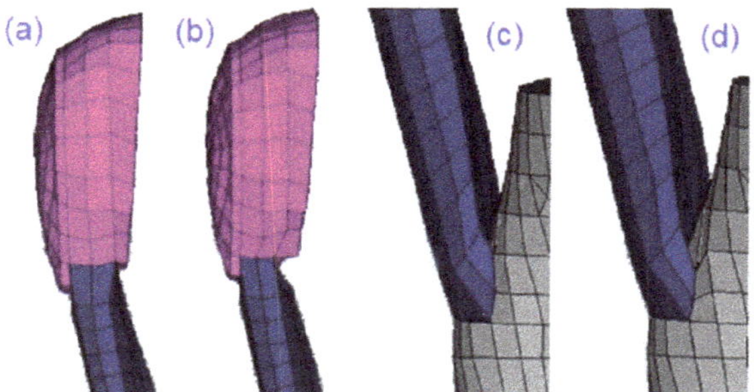

Figure 3. The finite element model of patellar tendon: (**a**) the upper before release; (**b**) the upper released; (**c**) the lower before release; (**d**) the lower released.

2.2. The Establishment of a Coordinate System for the Knee Joint's Motion

The corresponding reference coordinate systems of femur, tibia and patella are established by the following methods: (1) The femur posterior condylar cortical bone was contour-fitted to two laps, with one line through the center of two circles as the X axis of the femur. Parallel with lines of force and through the centerline of the femoral medial condyle represent the Z axis of the femur. Regarding the line, this passes through the center of the femoral medial condyle and perpendicular to the X-axis and Z-axis, as the Y-axis. (2) The intersection point of the Z-axis and the plane of the tibial joint line serves as the origin of the tibial coordinate system, and the femoral coordinate system moves to this point to obtain the tibial coordinate system. (3) The center of the patellar anterior surface moves inward 10 mm, then the center serves as the origin of the patellar coordinate system. The coordinate system of the femur was relocated to the origin, and the patellar coordinate system can be obtained [20]. The bone coordinate systems are shown in Figure 4.

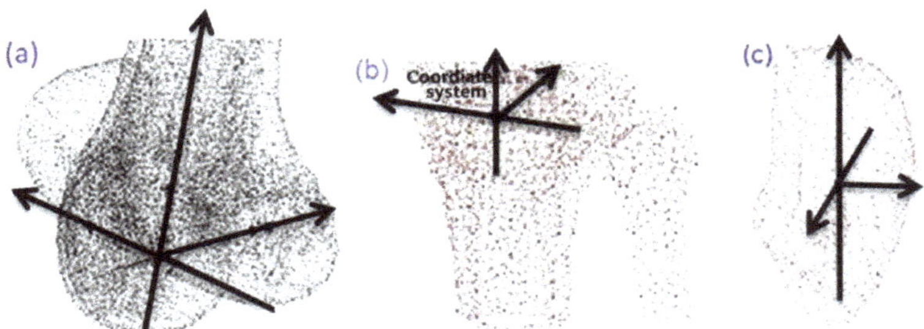

Figure 4. Coordinate system of bone. (**a**) femur; (**b**) tibia; (**c**) patella.

2.3. Loading Conditions and the Setting of Material Properties

For this study, 400 N [19,21] of force was applied to the quadriceps, and the direction of the force was parallel with the femoral shaft, pointing to the starting point of the quadriceps. At the same time, half of the body's weight (0.5BW) was applied along the force line of the knee joint, that is, a force of 300 N perpendicular to the ground was applied to the center of the femoral head. The boundary conditions were defined: the reference point of the ankle corresponding to the rotation center was fully fixed with six degrees of freedom. The material properties of bone tissue and femoral prosthesis were defined as isotropic and linear elastic, respectively. The material properties of polymer polyethylene pad were defined as non-linear elastic-plastic deformability [22,23]. The material properties of soft tissue were defined as nonlinear elastic [24]. The friction coefficient of polymer polyethylene and cobalt chromium molybdenum material was defined as 0.04 [22]. The penalty function with weighted factor was adopted [25,26]. For the femur, polymer polyethylene pad, tibial plateau, patellar prosthesis, and other soft tissues, seven surface contact pairs were defined in the finite element model after TKA. Moreover, nine surface contact pairs were defined in the finite element model of normal human.

3. Results

The four different knee joint finite element models were imported into the finite element analysis ABAQUS software, version 6.10 (Dassault SIMULIA Corp., Paris, France). Within 0–135 degrees, the data of motion characteristics at different flexion degrees were calculated and analyzed, as shown in Figure 5.

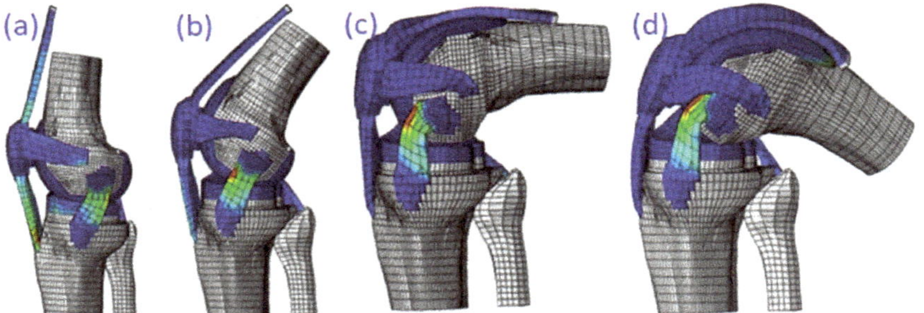

Figure 5. A sample of simulation results after TKA. Flexion degrees: (**a**) 0°; (**b**) 30°; (**c**) 90°; (**d**) 135°.

3.1. Medial-Lateral Shift and Flexion of Patella Along the Inner and Outer Axes

Figure 6a shows the medial-lateral shift data of the patellofemoral joint within 0–135 degrees of knee flexion. As the figure shows, from 0 to 30 degrees of flexion, the patella relative to the femur shifted outward. After releasing the patellar tendon, the patella shifted outward greatly. The lateral shift of the patellar tendon released at the lower end was larger than that of release at the upper end. The lateral shift was the largest when both ends were released simultaneously. From 40 degrees to 135 degrees of flexion, the patella relative to the femur shifted inward. After releasing the patellar tendon, the patella medial shift was smaller than that of no-release. The medial shift of the patellar tendon released at different ends was basically balanced. After the upper end of the patellar tendon was released, the maximum of the patella medial shift was 2.62 mm at 110 degrees of flexion, which was 11% lower than that of 2.94 mm at 100 degrees of flexion for the no-release model. After the lower end of the patellar tendon was released, the maximum medial shift of the patella was 2.63 mm at 110 degrees of flexion, and decreased 11% by compared with the no-release model. After both ends of the patellar tendon were released, the maximum medial shift of the patella was 2.61 mm at 110 degrees of flexion, and decreased 11% by compared with the no-release model.

Figure 6b shows the changes of patella flexion before and after the release. At the same flexion degree of the knee, the release of the patellar tendon at the upper end and lower end was compared with the no-release model, the flexion of the patella relative to femur was slightly larger. The patella flexion after both ends release (compared with the no-release model), in the same flexion degree of the knee, showed a smaller degree of patella flexion (within 130–135 degrees of flexion, being slightly larger than the no-release model). At 135 degrees of knee flexion, for the three releasing models: the upper end release, lower end release, and both ends release models, the patellar flexion reached maximums of 92, 92.73, and 93.17 degrees, respectively. Compared with the no-release model of 89 degrees, this shows an increase of 3%, 4%, and 5%, respectively.

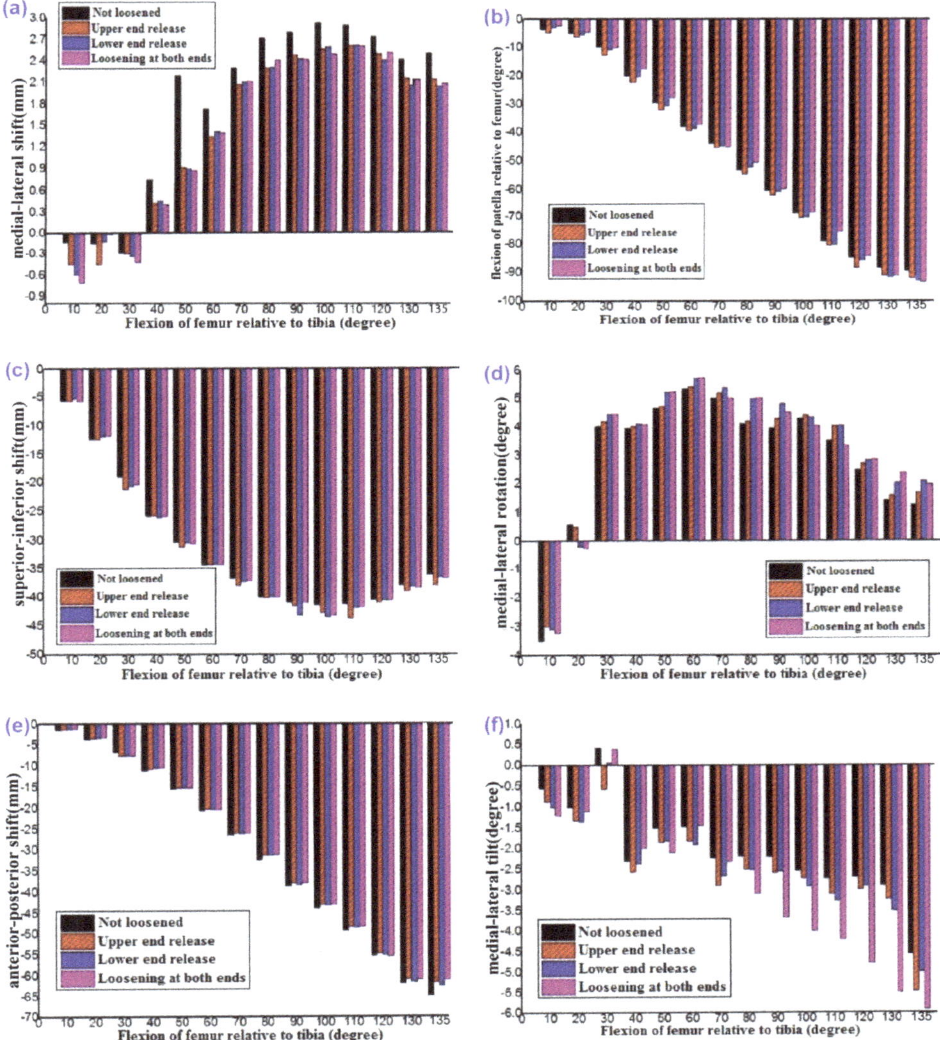

Figure 6. Patellofemoral movement data at different flexion degrees before and after the patellar tendon was released. (**a**) Medial-lateral shift; (**b**) flexion of patella relative to femur; (**c**) superior-inferior shift; (**d**) medial-lateral rotation; (**e**) anterior-posterior shift; (**f**) medial-lateral tilt.

3.2. Superior-Inferior Shift and Medial-Lateral Rotation of the Patella Along Its Upper and Lower Axes

As Figure 6c shows, during the knee joint's squatting movement, the patella relative to the femur shifted downward. The patella inferior shift of no-release model was slightly smaller than that of the upper end release. For the lower end release simulation, the patella inferior shift was slightly smaller within early 0–20 degrees of flexion and slightly larger after the 20-degree flexion than that of the no-release model. Within all flexion degrees, the patella inferior shift of the both ends release was slightly larger than that of no-release model, other than 20-degree flexion. After releasing the upper end of the patellar tendon, the maximum inferior shift of the patella was 43.80 mm at 110-degree flexion, which increased 6% compared with 41.42 mm at 100-degree flexion of the no-release model. For the lower end release model, the maximum inferior shift of the patella was 43.49 mm at 100-degree flexion, which increased 5% compared with the no-release model. After releasing the patellar tendon at both ends, the maximum inferior shift of the patellar was 43.43 mm at 100-degree flexion, which increased 5% compared with the no-release model.

Figure 6d shows the data of medial-lateral rotation. The data of the patella relative to the femur showed lateral rotation first and then medial rotation. Compared to the no-release model and the upper end model, the knee flexion degree of the initial patella medial rotation was greater for the lower end model and both-ends model. Meanwhile, it is shown that the patella medial-rotation for all three release models as larger than that of the no-release model. At 60 degrees of knee flexion, the patella medial-rotation reached the maximum of 5.41 degrees, 5.70 degrees, and 5.72 degrees, respectively, for the upper-end, lower-end and both-ends models, which increased 1%, 7%, and 7%, respectively, compared with the 5.34 degree value of the no-release model.

3.3. Anterior-Posterior Shift and Medial-Lateral Tilt of Patella Along the Anteroposterior Axis

Figure 6e shows the patella translated backward relative to the femur. During the knee joint's squatting movement, the patella relative to the femur translated backward. Within all flexion degrees, the patella backward translations of all three release models were smaller than that of no-release model, other than the 30-degree flexion, and the patella posterior shift was the least when both ends were released. At 135 degrees of knee flexion, the patella posterior shift reached the maximum values of 61.82 mm, 62.57 mm, and 61.07 mm, respectively, for the upper-end, lower-end, and both-ends models, which decreased 5%, 4%, and 6% compared with the value 64.97 mm of the no-release model.

Figure 6f shows the data of medial-lateral tilt. Before and after patellar tendon release, within all flexion degrees, the patella showed an inward movement pattern relative to the femur other than 30-degree flexion. At the same degree of knee flexion, after the release of the patella tendon, the medial tilt angle of the patella relative to the femur was larger than that of no release. After 80 degrees of knee flexion, when both ends of the patellar tendon were released at the same time, the patella medial tilt angle was the largest. At 135 degrees of knee flexion, the patella medial tilt angle reached the maximum values of 5.51 degrees, 5.01 degrees, and 5.92 degrees, respectively, for the upper-end, lower-end, and both-ends models, which increased 21%, 10%, and 30%, respectively, compared with the 4.55 degree value of the no-release model.

4. Discussion

This paper investigates the patellar tendon release's effect on the movement characteristics of the artificial patellofemoral joint squat. The simulation results show that, after the patellar tendon being released, respectively, at the upper end, the lower end, and both ends, being compared with the no-release simulation results, the relative flexion, medial-lateral rotation, medial-lateral tilt, and superior-inferior shift of the patella relative to the femur increased, and the medial-lateral shift and anterior-posterior shift of the patella relative to the femur decreased. Different release models have different effects on the squatting motion characteristics of the patellofemoral joint. In this study, the data of patellofemoral joint squatting after TKA of the no-release model was analyzed and compared with those of all three release models and other related research results of the natural knee joint.

4.1. Medial-Lateral Shift and Flexion of the Patella

For the medial-lateral shift of the patella relative to the femur the maximum patella medial-shift of all three release models were balanced, and the maximum medial-shift was decreased compared with that of the no-release model (the maximum reduction was about 11% for the both-ends model). The patella shifted outward first and then inward relative to the femur, which is similar to the related studies [27,28].

For the flexion movement of the patella, relative to the femur, previous studies [27,28] showed that the flexion of patella changes linearly with the flexion of the patellofemoral joint. It is consistent with the results of this study. In this study, the maximum flexion angle of the patella increased after the patellar tendon being released (compared with the no-release model, the maximum of the both-ends model increased by 5%), which indicated that the mobility of the knee joint was improved after the patellar tendon release.

4.2. Superior-Inferior Shift and Medial-Lateral Rotation of Patella

For the superior-inferior shift of the patella, relative to the femur, after the patellar tendon was released in different ends, the patella shifted downward relative to the femur. In the high flexion, the patella was moved up slightly, and the maximum inferior shift (about 43.5 mm at 110 degrees) was increased 5% compared with that of no-release model. The maximum increase occurred in the upper-end release model and the minimum one occurred at both-ends release model. Related studies [27,29] also showed that the patella, with the knee flexion, shifted downward relative to the femur, and the maximum flexion angle of the knee was 110 degrees in their studies (the maximum inferior shift was about 50 mm), which was basically similar to this paper (in this paper, the maximum flexion angle of the knee was 135 degrees, and the maximum inferior shift was 43.8 mm).

For the medial-lateral rotation of the patella, relative to the femur, after the patellar tendon was released at different ends, the motion pattern of the patella relative to the femur was lateral rotation first and then medial rotation, which is consistent with that of the no-release model. For all the above three release models, the maximum medial rotation of the patella increased; among them, the increase of the upper end release was the minimum (1%), and the increase of the both-ends release were the maximum (7%). Heegaard et al. [30] showed that within 0–110 degrees of knee flexion, the patellar rotation pattern changed from medial to lateral rotation, and then lateral rotation within 110–135 degrees of knee flexion. This phenomenon indicates that the patellar motion changes during the high flexion, which might be related to the way of femoral fixation and passive loading.

4.3. Anterior-Posterior Shift and Medial-Lateral Tilt of the Patella

For the anterior-posterior shift of the patella, relative to the femur, the results of Azmy et al.'s [31] in vitro experiments showed that with the flexion of the knee joint, the patella shifted backward; after 30 degree knee flexion, posterior movement rate of the patella increased, and the maximum posterior movement of the patella reached 32 mm. The results of Baldwin et al. [29], based on the experimental dynamic finite element model, also show that the patella continues to shift backward with the knee flexion, and the maximum posterior movement of the patella is about 30 mm. The motion analysis results of patella anterior-posterior shift in this paper are consistent with those of the above studies; with knee flexion, patella continued to shift backward. Meanwhile, the analysis results of this paper gave the data of the posterior shift of the patella during high knee flexion. At 135 degree knee flexion, the posterior shift of the patella after the upper end, the lower end, and both ends of the patellar tendon were released to reach the maximum values of 61.82 mm, 62.57 mm, and 61.07 mm, respectively, which was 5%, 4%, and 6% lower than the no-release values of 64.97 mm.

For the medial-lateral tilt of the patella relative to the femur with the flexion of the knee joint, the patella relative to the femur exhibits an inward motion pattern before and after release. At the same degree of flexion, the medial tilt angle of the patella relative to the femur was larger than that of the no-release model, the increase of the both ends release was the maximum (30%), and the increase of the lower end release was the minimum (10%). The relevant findings [27,29,31,32] also showed that the patella, with the flexion of knee, tilted inward relative to the femur, which is consistent with the results of this study. At the same time, there are different research results for the medial-lateral tilt motion of the patella [33]. The results of Amis et al. [33] showed that the patella continued to tilt outward with knee flexion, which may be related to the loading method that the quadriceps were divided into six bundles. There are differences in related studies, which may be related to the fact that the patella was less constrained by the femoral articular surface in the direction of patella medial-lateral tilt.

Several limitations of the present study should be noted. In this study, only 30% of the patella was released, and others were not analyzed. The finite element model of the human knee joint after TKA was established in this paper, based on the consideration of all bone tissue and the main soft tissue of the knee; although the transverse patellar ligaments, which play an important role in the movement of the patella, are added, the effect of the joint capsule, hamstring, or gastrocnemius on knee joint activity was not considered. In the subsequent analysis of the knee joint, these tissues could be taken into account to make the model more in line with human physiology.

The anterior cruciate ligament (ACL) has the functions of restricting the forward movement of the tibia and controlling the internal and external valgus of the knee [34–36]. The ACL is extremely vulnerable during exercise and it is difficult to repair itself. If not treated in time, the patient's daily activities will be seriously affected [37]. The posterior cruciate ligament (PCL), aside from providing restraint to posterior tibial translation, is reported to have a considerable role in providing rotational stability to the knee [38]. Injuries to the PCL are far less common than injuries to other knee structures, such as the ACL or the menisci [39]. In addition, it has been found that the clinical outcomes of PCL reconstruction are less satisfactory and less predictable relative to those of ACL reconstruction [39]. The medial patellofemoral ligament (MPFL) is one of the most important restraints to lateral displacement of the patella from 0° to 30° of flexion, providing up to 60% of lateral patellofemoral stability [40]. In the last decade, the understanding of the pathoanatomic mechanisms involved in lateral patellar dislocation (LPD) has evolved tremendously and reconstruction of the medial patellofemoral ligament (MPFL), as the most important passive restraint against LPD, has evolved to become an established operative procedure [41]. This paper mainly investigates the patellar tendon release's effect on the movement characteristics of the artificial patellofemoral joint squat. However, different knee ligaments play different roles. Based on the models used in this study, we will further study the ligaments' reconstruction and other related issues.

5. Conclusions

This paper investigates the patellar tendon release's effect on the movement characteristics of the artificial patellofemoral joint squat. The post-TKA patellofemoral joint's motion characteristics in different patellar tendon release pattern were obtained. Compared with the no-release model, the flexion, medial-lateral rotation, medial-lateral tilt, and superior-inferior shift of the patella relative to the femur increased; and the medial-lateral shift and anterior-posterior shift of the patella relative to the femur decreased. In this paper, the maximum flexion angle of the patella increased after the patellar tendon being released (compared with the no-release model), which indicated that the mobility of knee joint was improved after the patellar tendon release. Different release models have different effects on the squatting motion characteristics of patellofemoral joint. In this paper, the simulation results can be used to moderately release the ends of the patellar tendon in the TKA surgery, in order to improve the motility of the patella relative to the femur. The results of this paper can provide a reference for the study of the pathology and rehabilitation of the knee joint, as well as related operations.

Author Contributions: Conceptualization: J.W.; data curation: L.F. and Y.L.; investigation: J.W. and L.F.; methodology: J.W. and Y.L.; resources: J.W.; software: J.W., Y.Y., and D.G.; supervision: J.W. and Y.L.; validation: Y.Y., D.G., and S.W.; writing—original draft: J.W.; writing—review and editing: Y.Y., D.G., S.W. and Y.L.

Funding: This research was funded by National Natural Science Foundation of China, grant number 31370999.

Conflicts of Interest: The authors declare no conflict of interest.

References

1. Adravanti, P.; Tecame, A.; De Girolamo, L.; Ampollini, A.; Papalia, R. Patella Resurfacing in Total Knee Arthroplasty: A Series of 1280 Patients at Midterm Follow-Up. *J. Arthroplast.* **2018**, *33*, 696–699. [CrossRef] [PubMed]
2. Chonko, D.J.; Lombardi, A.V.; Berend, K.R. Patella baja and total knee arthroplasty (TKA): Etiology, diagnosis, and management. *Surg. Technol. Int.* **2004**, *12*, 231–238. [PubMed]
3. Ali, S.A.; Helmer, R.; Terk, M.R. Patella Alta: Lack of Correlation Between Patellotrochlear Cartilage Congruence and Commonly Used Patellar Height Ratios. *J. Am. Roentgenol.* **2009**, *193*, 1361–1366. [CrossRef] [PubMed]
4. Nakagawa, S.; Arai, Y.; Inoue, H.; Atsumi, S.; Ichimaru, S.; Ikoma, K.; Fujiwara, H.; Kubo, T. Two Patients with Osteochondral Injury of the Weight-Bearing Portion of the Lateral Femoral Condyle Associated with Lateral Dislocation of the Patella. *Case Rep. Orthop.* **2014**, *2014*, 876410. [CrossRef] [PubMed]
5. Chougule, S.S.; Stefanakis, G.; Stefan, S.C.; Rudra, S.; Tselentakis, G. Effects of fat pad excision on length of the patellar tendon after total knee replacement. *J. Orthop.* **2015**, *12*, 197–204. [CrossRef]
6. Weale, A.E.; Murray, D.W.; Newman, J.H.; Ackroyd, C.E. The length of the patellar tendon after unicompartmental and total knee replacement. *Bone Jt. J.* **1999**, *81*, 790–795. [CrossRef]
7. Davies, G.S.; Van Duren, B.; Shorthose, M.; Roberts, P.G.; Morley, J.R.; Monk, A.P. Changes in patella tendon length over 5 years after different types of knee arthroplasty. *Knee Surg. Sports Traumatol. Arthrosc.* **2016**, *24*, 3029–3035. [CrossRef]
8. FlORen, M.; Davis, J.; Peterson MG, E.; Laskin, R.S. A mini-midvastus capsular approach with patellar displacement decreases the prevalence of patella baja. *J. Arthroplast.* **2007**, *22*, 51–57. [CrossRef]
9. Meneghini, R.M.; Ritter, M.A.; Pierson, J.L.; Meding, J.B.; Berend, M.E.; Faris, P.M. The effect of the Insall-Salvati ratio on outcome after total knee arthroplasty. *J. Arthroplast.* **2006**, *21*, 116–120. [CrossRef]
10. Kazemi, S.M.; Daftari, B.L.; Eajazi, A.; Miniator Sajadi, M.R.; Okhovatpoor, M.A.; Farhang, Z.R. Pseudo-patella baja after total knee arthroplasty. *Med. Sci. Monit. Int. Med. J. Exp. Clin. Res.* **2011**, *17*, CR292. [CrossRef]
11. Krishnan, S.G.; Steadman, J.R.; Millett, P.J.; Hydeman, K.; Close, M. *Lysis of Pretibial Patellar Tendon Adhesions (Anterior Interval Release) to Treat Anterior Knee Pain After ACL Reconstruction*; Springer: London, UK, 2006.
12. Seo, J.G.; Moon, Y.W.; Kim, S.M.; Park, S.H.; Lee, B.H.; Chang, M.J.; Jo, B.C. Prevention of pseudo-patella baja during total knee arthroplasty. *Knee Surg. Sports Traumatol. Arthrosc.* **2015**, *23*, 3601–3606. [CrossRef] [PubMed]
13. Bugelli, G.; Ascione, F.; Cazzella, N.; Franceschetti, E.; Franceschi, F.; Dell'Osso, G.; Giannotti, S. Pseudo-patella baja: A minor yet frequent complication of total knee arthroplasty. *Knee Surg. Sports Traumatol. Arthrosc.* **2017**, *26*, 1831–1837. [CrossRef] [PubMed]
14. Wang, J.-P.; Qian, L.-W.; Wang, C.-T. Simulation of Geometric Anatomy Model of Human Knee Joint. *J. Syst. Simul.* **2009**, *21*, 2806–2809.
15. Matava, M.J.; Hutton, W.C. A biomechanical comparison between the central one-third patellar tendon and the residual tendon. *Br. J. Sports Med.* **1995**, *29*, 178–184. [CrossRef]
16. Mansson, O.; Sernert, N.; Rostgard-Christensen, L.; Kartus, J. Long-term clinical and radiographic results after delayed anterior cruciate ligament reconstruction in adolescents. *Am. J. Sports Med.* **2015**, *43*, 138–145. [CrossRef]
17. George, D.S.L.M.; Pampolha, A.G.M.; Orlando Junior, N. Functional results from reconstruction of the anterior cruciate ligament using the central third of the patellar ligament and flexor tendons. *Rev. Bras. Ortop. (Engl. Ed.)* **2015**, *50*, 705–711.
18. Haviv, B.; Yassin, M.; Rath, E.; Bronak, S. Prevalence and clinical implications of nerve injury during bone patellar tendon bone harvesting for anterior cruciate ligament reconstruction. *J. Orthop. Surg.* **2017**, *25*, 230949901668498. [CrossRef]

19. Wang, J.; Tao, K.; Li, H.; Wang, C. Modelling and analysis on biomechanical dynamic characteristics of knee flexion movement under squatting. *Sci. World J.* **2014**, *2014*, 321080.
20. Grood, E.S.; Suntay, W.J. A joint coordinate system for the clinical description of three-dimensional motions: Application to the knee. *J. Biomech. Eng.* **1983**, *105*, 136–144. [CrossRef]
21. Li, G.; Most, E.; Otterberg, E.; Sabbag, K.; Zayontz, S.; Johnson, T.; Rubash, H. Biomechanics of Posterior-Substituting Total Knee Arthroplasty: An In Vitro Study. *Clin. Orthop. Relat. Res.* **2002**, *404*, 214–225. [CrossRef]
22. Godest, A.C.; Beaugonin, M.; Haug, E.; Taylor, M. Simulation of a knee joint replacement during a gait cycle using explicit finite element analysis. *J. Biomech.* **2002**, *35*, 267–275. [CrossRef]
23. Taylor, M.; Barrett, D.S. Explicit finite element simulation of eccentric loading in total knee replacement. *Clin. Orthop. Relat. Rec* **2003**, *414*, 162–171. [CrossRef] [PubMed]
24. Pena, E.; Calvo, B.; Martinez, M.A.; Doblare, M. A three-dimensional finite element analysis of the combined behavior of ligaments and menisci in the healthy human knee joint. *J. Biomech.* **2006**, *39*, 1686–1701. [CrossRef] [PubMed]
25. Halloran, J.P.; Petrella, A.J.; Rullkoetter, P.J. Explicit finite element modeling of total knee replacement mechanics. *J. Biomech.* **2005**, *38*, 323–331. [CrossRef]
26. Bustince, H.; Fernandez, J.; Burillo, P. *Penalty Function in Optimization Problems: A Review of Recent Developments*; Springer: Cham, Switzerland, 2018.
27. Tang, T.S.; MacIntyre, N.J.; Gill, H.S.; Fellows, R.A.; Hill, N.A.; Wilson, D.R. Accurate assessment of patellar tracking using fiducial and intensity-based fluoroscopic techniques. *Med. Image Anal.* **2004**, *8*, 343–351. [CrossRef] [PubMed]
28. Fellows, R.A.; Hill, N.A.; Gill, H.S.; MacIntyre, N.J.; Harrison, M.M.; Ellis, R.E.; Wilson, D.R. Magnetic resonance imaging for in vivo assessment of three-dimensional patellar tracking. *J. Biomech.* **2005**, *38*, 1643–1652. [CrossRef]
29. Baldwin, M.A.; Clary, C.; Maletsky, L.P.; Rullkoetter, P.J. Verification of predicted specimen-specific natural and implanted patellofemoral kinematics during simulated deep knee bend. *J. Biomech.* **2009**, *42*, 2341–2348. [CrossRef]
30. Heegaard, J.; Leyvraz, P.F.; Curnier, A.; Rakotomanana, L.; Huiskes, R. The biomechanics of the human patella during passive knee flexion. *J. Biomech.* **1995**, *28*, 1265–1279. [CrossRef]
31. Azmy, C.; Guerard, S.; Bonnet, X.; Gabrielli, F.; Skalli, W. EOS (R) orthopaedic imaging system to study patellofemoral kinematics: Assessment of uncertainty. *Orthop. Traumatol. Surg. Res.* **2010**, *96*, 28–36. [CrossRef]
32. McWalter, E.J.; Hunter, D.J.; Wilson, D.R. The effect of load magnitude on three-dimensional patellar kinematics in vivo. *J. Biomech.* **2010**, *43*, 1890–1897. [CrossRef]
33. Amis, A.A.; Senavongse, W.; Bull, A.M.J. Patellofemoral kinematics during knee flexion-extension: An in vitro study. *J. Orthop. Res.* **2006**, *24*, 2201–2211. [CrossRef] [PubMed]
34. Testa, R. Knee rotational laxity and proprioceptive function 2A years after partial ACL reconstruction. *Knee Surg. Sports Traumatol. Arthrosc.* **2012**, *20*, 762–766.
35. Gokeler, A.; Benjaminse, A.; Hewett, T.E.; Lephart, S.M.; Engebretsen, L.; Ageberg, E.; Dijkstra, P.U. Proprioceptive deficits after ACL injury: Are they clinically relevant? *Br. J. Sports Med.* **2012**, *46*, 180–192. [CrossRef] [PubMed]
36. Smith, C.K.; Howell, S.M.; Hull, M.L. Anterior laxity, slippage, and recovery of function in the first year after tibialis allograft anterior cruciate ligament reconstruction. *Am. J. Sports Med.* **2011**, *39*, 78. [CrossRef]
37. Tianqu, H. Finite Element Analysis of the Influence of Femoral Tunnel Locating Points on the Isometricity of Grafts in Anterior Cruciate Ligament Reconstruction. Master's Thesis, Central South University, Changsha, China, 24 May 2014.
38. Laprade, C.M.; Civitarese, D.M.; Rasmussen, M.T.; Laprade, R.F. Emerging Updates on the Posterior Cruciate Ligament: A Review of the Current Literature. *Am. J. Sports Med.* **2015**, 0363546515572770. [CrossRef]
39. Narvy, S.J.; Pearl, M.; Vrla, M.; Yi, A.; Hatch, G.F.R. Anatomy of the Femoral Footprint of the Posterior Cruciate Ligament: A Systematic Review. *Arthrosc. J. Arthrosc. Relat. Surg.* **2015**, *31*, 345–354. [CrossRef]

40. Tucker, A.; Mcmahon, S.; Mcardle, B.; Rutherford, B.; Acton, D. Synthetic versus autologous reconstruction (Syn-VAR) of the medial patellofemoral ligament: A study protocol for a randomised controlled trial. *Trials* **2018**, *19*, 268. [CrossRef]
41. Balcarek, P.; Rehn, S.; Howells, N.R.; Eldridge, J.D.; Kita, K.; Dejour, D.; Friede, T. Results of medial patellofemoral ligament reconstruction compared with trochleoplasty plus individual extensor apparatus balancing in patellar instability caused by severe trochlear dysplasia: A systematic review and meta-analysis. *Knee Surg. Sports Traumatol. Arthrosc.* **2016**, *25*, 3869–3877. [CrossRef]

Article

User Friendliness of a Wearable Visual Behavior Monitor for Cataract and Refractive Surgery

Bojan Pajic [1,2,3,4,*], Pavel Zakharov [5], Brigitte Pajic-Eggspuehler [1] and Zeljka Cvejic [2]

1   Eye Clinic Orasis, Swiss Eye Research Foundation, Titlisstrasse 44, 5734 Reinach AG, Switzerland; brigitte.pajic@orasis.ch
2   Department of Physics, Faculty of Sciences, University of Novi Sad, Trg Dositeja Obradovica 4, 21000 Novi Sad, Serbia; zeljka.cvejic@df.uns.ac.rs
3   Division of Ophthalmology, Department of Clinical Neurosciences, Geneva University Hospitals, 1205 Geneva, Switzerland
4   Faculty of Medicine of the Military Medical Academy, University of Defense, 11000 Belgrade, Serbia
5   Vivior AG, Technoparkstrasse 1, 8005 Zürich, Switzerland; pavel.zakharov@vivior.com
*   Correspondence: bpajic@datacomm.ch; Tel.: +41-62-765-6080

Received: 23 December 2019; Accepted: 17 March 2020; Published: 24 March 2020

Abstract: A prospective feasibility study was conducted to determine whether a new wearable device, the Visual Behavior Monitor (VBM), was easy to use and did not present any difficulties with the daily activities of patients. Patients for cataract surgery and refractive lens exchange were randomly selected and screened for inclusion in the study. A total of 129 patients were included in the study as part of a multicenter study. All measurements were performed before surgery. Upon inclusion, patients were trained to wear the device, instructed to wear it for a minimum of 36 h, and were scheduled to return in one week. The VBM measures the distance at which patients' visual activities are performed, the level of illumination, and head translational and rotational movements along the three axes. On the follow-up visit, patients completed a questionnaire about their experience in wearing the device. All patients underwent standard diagnostic testing, with their cataract grade determined by the Lens Opacities Classification System (LOCS) classification. Results indicate that 87% of patients felt comfortable using the wearable device while 8% of patients responded as not feeling comfortable (5% of patients did not respond to the question). In addition, 91% of patients found it easy to attach the wearable to the magnetic clip while 4% of patients did not find it easy, and 5% of patients did not respond. Overall, patients found the device easy to use, with most reporting that the device was not intrusive.

Keywords: cataract surgery; refractive surgery; visual behavior monitor

---

1. Introduction

Cataract surgery has become the most common operation worldwide. In the United States alone, 3.6 million cataract surgeries are performed each year. This corresponds to a cataract surgery rate (CSR) of 11,000 per 1 million inhabitants. In economically well-developed countries such as the United States, Europe, Australia, or Japan, the CSR is between 4000 and 11,000 [1,2]. Yet, refractive outcomes, and patient satisfaction, following cataract surgery remain a key challenge for ophthalmologists today, particularly when it involves a presbyopia-correcting intraocular lens (IOL) [3].

In an era where we strive to improve on our outcomes by using increasingly sophisticated biometry and IOL power calculations, it is important not to lose sight of the impact of residual or increased refractive errors that may not be apparent using the spherical equivalent or viewing the components of the refractive error in isolation on the patient's vision. We should use the tools that facilitate surgeons to assess and manage the patient's refractive error in its entirety.

Approaching the intended post-operative spherical equivalent, however, often does not achieve spectacle independence. Uncorrected residual spherocylindrical refractive errors appear to have far greater adverse effects on unaided visual acuity than may be evident using a spherical equivalent or the individual sphere and cylinder. If the correction of presbyopia is desired, there are three options available for IOLs in the course of cataract surgery. One is the installation of a monovision where the dominant eye is set for distance and the non-dominant eye for near. A difference of up to 2 D is accepted by the patient. However, before proceeding to surgery, it is essential to try wearing contact lenses to see if the correction is tolerated [4,5]. Most patients reported the surgery met their expectations for decreased dependence on spectacles (93%) [6]. The extended-depth-of-focus-IOL (EDOF-IOL) provides an extended depth-of-field range. Actually, it is a bifocal IOL with a low near-addition of 1.5–2.0 D. This results in excellent distance and intermediate correction without the disadvantages of a multifocal IOL. However, this procedure does not completely eliminate the need for spectacles [5,7]. The third possibility of correction is the implantation of a trifocal IOL. One is located in the distance and receives 50% of the incident light, the second is at 30–40 cm and receives 30% of the light, and the third is at about 60 cm and receives 20% of the light. The strength of trifocal IOLs is that they are glasses-free. The weakness is that fixed distances are given, which requires some getting used to, and that the intermediate part is the weakest. As only a part of the light used for image generation is sharp, there is a loss of contrast, which is noticeable. In addition, undesirable optical side effects in the form of so-called halos or glare occur, which impair vision, especially at twilight and at night [5,8]. There is no ideal correction for all distances, but through objective measurement using the Visual Behavior Monitor (VBM) regarding distance measurement, ambient lighting and head posture will potentially determine a much more accurate setting and support the choice of IOL. Whatever additional technique is used, an important measure of success is the difference between the intended and the actual postoperative refractive outcome. It is necessary to have a method that is sensitive to discrepancies between the intended and the actual outcome.

Hence, to minimize potential side effects and improve patient outcomes, IOLs must be chosen with the visual needs of a patient in mind. One such method to assess a patient's individual visual needs is to determine patient expectations and visual behavior pre-operatively, to improve post-operative outcomes.

Visual behavior monitors are developed through various technological approaches, which should be as precise as possible but also cost-effective. Various studies use active lighting based on infrared LEDs. For example, one study [9] proposes a system that uses 3D vision techniques to estimate and track the 3D line of sight of a person with multiple cameras. The method is based on a simplified eye model and uses the Purkinje images of an infrared light source to determine the eye position. This information is used to estimate the line of sight. In a technological development, a system with active infrared LED illumination and one camera is implemented. Due to the LED illumination, the method can easily find the eyes and the system uses this information to locate the remaining facial features [10]. The authors suggest estimating the local direction of gaze analytically based on the pupil position. Almost all active systems described in the literature have been tested in simulated environments, but not in real ones. A moving person presents new challenges such as variable lighting, changing background, and vibrations that have to be considered in real systems. In a further work [11], an industrial prototype called Copilot is presented whose application tends to take into account the real situation. This system uses an infrared LED illumination to find the eyes. It uses a simple subtraction procedure to find the eyes and calculates only one validated parameter, the percentage eye closure (PERCLOS). This system currently works in low-light conditions [12]. All of these technologies, including ours, tend to collect even more precise data and reflect reality. Further improvements can be expected in the future by combining different technologies. Many technologies are developed in such a way that they are easy to handle and can be spread further. For example, the investigation tools are built into smartphones [13]. This is the way we initially went. However, a further simplification of the operation has become necessary after initial experiences. This led to today's VBM, which does not require any technical knowledge of the patient. Due to the simplicity, the measurement results are much more precise and reflect reality much better.

Results from a literature review suggest that there is a greater utility of using measurable patient reported outcomes such as functioning, satisfaction, and quality of life as opposed to clinical outcomes when considering the benefits of cataract surgery [1,14].

Psychometric tests in the form of cataract surgery outcome questionnaires have been developed, identifying the limitations of performing daily tasks due to an individual's vision, a trait known as visual functioning [15,16]. Analyzing the visual behavior of cataract patients is extremely important for defining the visual strategy adopted in the patient's care.

McAlinden et al. [16] studied the responsiveness of several questionnaires in measuring patient outcomes. The investigation revealed that "Catquest-9SF," a Rasch modified 9 item questionnaire originally developed in Sweden [17], was the most responsive as measured by an effect size (ES) statistic of 1.45 (95% CI, 1.22–1.67). Although the "Catquest-9SF" includes a measure of visual functioning, a questionnaire known as the "Visual Disability Assessment (VDA)" (ES, 1.09; 95% CI, 0.86–1.32) also includes a measure of mobility of the patient, which may be attractive for patients and clinicians. Finally, the VBM measurement supports the solution finding for a planned surgery.

Morlock et al. [18] developed a Patient-Reported Spectacle Independence Questionnaire (PRSIQ) to assess spectacle independence following cataract surgery. The questionnaire was developed using conventional qualitative and modern measurement theory methods, and evidence was garnered for the use of PRSIQ as a measure of spectacle independence.

While the use of questionnaires and visual assessments in measuring patient outcomes are effective, there are limitations regarding the subjectivity of the results. Current psychometric measures may be subject to systematic error such as recall bias and could cause judgment errors resulting in patient discontentment with surgical outcomes [19,20]. Visual information based on patient behavior can be harnessed using technological tools that perform eye tracking, analyzing visual behavior under various circumstances. Eye tracking technologies can be used to attain and process objective information of a patient's everyday activities, crucial for selecting an appropriate IOL.

Integrating technology that collects objective visual behavior data into pre-operative patient assessments has shown to improve the current understanding of patient behavioral gaze. The data generated could aid clinicians to provide a highly individual evaluation of a patient's visual needs, including distance estimates of the patient's sight, and determine the right course of action to improve post-operative patient outcomes and patient satisfaction.

To date, most approaches have focused on diagnostic testing in a clinical setting. In-clinic measurements cannot fully replicate a patient's daily visual needs. The Visual Behavior Monitor (VBM, Vivior AG, Zurich, Switzerland) is a wearable device to be attached on the spectacles of a patient and perform continuous measurements as he or she goes about their daily activities. The challenge of unassisted monitoring in outpatient settings is ensuring correct use of the monitoring system and reliability of the data. This is especially the case for the senior cataract patients with compromised visual function due to age-related cataract and/or presbyopia. While aiming to develop a compact and non-obtrusive system, it is essential to ensure that the reduced form-factor of the device does not impede the handling of the device by the elderly population. Another challenge is to design the user interfaces of the system in a way that can be correctly interpreted by patients with different levels of exposure to computer and information technologies. Most of the recent progress in the information technologies has been pushed forward generally by the young generations of earlier adopters and, thus, interfaces are shaped for and by this population group, who have enough time and interest to work with new technologies. It is critically important to redesign the user experience from scratch with wearable systems for all population groups.

Inadequately designed systems ignoring the specifics of the population groups might not only be inefficient in delivering required objective information, but also lead to reduced patient satisfaction due to frustration from handling the device. This is aggravated by the patient awareness that the outcome of the surgery might depend on his or her ability to handle the system. With all the above in mind, the system should be comfortable for use by the patients of various age groups, educational backgrounds with different levels of exposure to information technology, and possibly with compromised vision.

The present study was initiated to understand the feasibility of the non-intrusive monitoring of visual behavior and VBM handling by, particularly, a cataract-age population without supervision of a healthcare professional (HCP). The study is designed to evaluate the ease of use, as well as patient perception of wearing the device. This can potentially answer the question of whether patients are willing to use and able to handle the wearable device mounted on the spectacles and are comfortable to use it in order to perform the required measurements.

2. Materials and Methods

Patients presenting for cataract surgery or refractive lens exchange were randomly selected to take part in this feasibility study to evaluate the VBM. Patients who met the inclusion criteria were recruited consecutively for the study. Upon enrolment, patients completed a modified Cat-Quest 9 questionnaire. For German-speaking sites, the questionnaire was modified to improve readability in German. Patients were trained and fitted with the VBM and instructed to wear the device for at least 36 h. Patients then returned a week later to complete a questionnaire on the user friendliness of the device and schedule surgery. In our study, we used a modified CATQ-9 questionnaire, which refers to the patient's ability to perform certain tasks. The questionnaire consists of 9 questions, with each question being rated on a scale of 1–4—from 1 for "no problems" to 4 for "very big problems." In general, it is asked whether the patient has problems with vision in everyday life and how satisfied the patient is with his or her vision. In particular, the questions ask how well the patient sees on the computer, how well he/she recognizes faces, how well he/she recognizes prices when shopping, how well he/she walks on uneven ground, how he/she watches television with subtitles, and what is his/her favorite pastime and hobbies. This covers a large part of daily life.

The following inclusion criteria are applied: Age over 18 years old, participants considering IOL procedures, participants considering refractive laser procedures, and participants willing and able to participate. As exclusion criteria, we have defined the following points for the study: Unwillingness to participate, manifestation of macular disease based on OCT(Optical Coherence Tomography) /fundus photography, corneal scar or sign of corneal decompensation, standard surgery exclusion criteria, inability to make all postoperative visits, history or manifestation of other pre-existing ocular conditions and comorbidity (amblyopia, monophthalmia), other predisposing sight-threatening ocular conditions (uveitis, diabetic retinopathy, ARMD, macular dystrophy, retinal detachment, neuro-ophthalmic disease, abnormal pupils [deformation] and iris), history of corneal or retinal surgery, evidence of iris or chorioretinal neovascularization, history of anterior segment pathology (aniridia), any scar from corneal endothelium damage (chemical burn, herpetic keratitis, cornea guttata), any medical history contraindicative of standard cataract surgery including those taking medications known to potentially complicate cataract surgery (e.g., α1a-selective alpha blocker), and participating in another clinical trial.

The primary endpoint of the study is testing the design and usability of the wearable and the station.

The Visual Behavior Monitor is a wearable device to continuously monitor visual behavior of the user. During the study, two functionally equivalent versions of the system were used, called Mark II and Mark III. The description below focuses on the latest system, Mark III.

The VBM is fixed on the spectacles frame with help of an adapter. Multiple adapters are provided to patients to allow attachment on all spectacles used by the patient. Patients can easily move the device from the adapter on one spectacle to the adapter on another when changing spectacles. The VBM is configured to start and stop measurements automatically when attached and detached from the adapter, as well as when spectacles are taken off.

After the VBM measurement was carried out, the respondent was asked the following questions: 1. Is it easy to attach the magnetic clip to the spectacles frames? 2. Is it easy is to attach the wearable to the magnetic clip? 3. Do you always feel comfortable using the wearable? 4. Is it easy to use the wearable?

Regarding the station, the following questions were asked. 5. Is the connection between the station and the wearable always working? 6. Is the information provided on screen always clear to you?

VBM (Figure 1) incorporates the following sensors: Two optical time-of-flight distance sensors directed forward and 30° downward, ambient light sensor with separate red, green, and blue channels directed forward, combined ultraviolet and ambient light sensor directed upward, as well as motion and orientation sensors: Accelerometer, gyroscope, and magnetometer. The VBM does not include imaging sensors, such as cameras and location sensors, and does not infringe on the privacy of the wearer and other people. The VBM features an energy-independent real-time clock for reliable timestamping of the measurements.

Figure 1. Visual Behavior Monitor (VBM) fixed on the spectacles.

Measurements data are stored in the encrypted format in the device and the progress of data collection is shown on the embedded screen or on the screen of the station in hours and minutes. The onboard memory capacity is more than 1000 h. The VBM is charged overnight and the battery capacity is sufficient for a complete day of continuous measurements. When the device is returned to the clinic, data are transferred from the device to an electronic tablet and then uploaded to the internet server (cloud), decrypted, processed, and visualized via web-interface.

Cloud processing is performed by advanced machine learning algorithms, which are trained to recognize various types of patient's visual activities, such as looking on a desktop computer screen, reading handheld materials, driving a car, and viewing objects at a distance. The environment is further analyzed based on the recognized activities to derive relevant metrics such as viewing distances to the objects of visual activities, such as distance to a computer screen, to a desktop, or to handheld material. The data are further aggregated to create relevant statistical representations.

The VBM is intended to provide behavioral information to the attending HCP and to support ophthalmic surgeons or other healthcare professionals in the planning of cataract and refractive surgery. This is expected to improve participant satisfaction through the objective evaluation of individual vision needs and allowing practitioners to more adequately address those needs. Based on the individual profile of viewing distances, it is possible to select the optimal IOL solution by overlaying the defocus curve of the IOL, representing the expected visual acuity as a function of defocus (visual distance). Additional measured parameters, such as typical light levels and their spectral content, can be also considered. For example, monofocal IOLs deliver good far vision but cannot provide clear vision in near and intermediate distance zones, while lenses with advanced optics, such as bifocal and trifocal (multifocal) lenses, as well as lenses with an extended depth of focus, are capable of providing sufficient visual acuity and spectacle independence both in the near and intermediate zones, depending on the lens. At the same time, advanced optics can cause halos and glare in low-light conditions. Thus, for a patient with far distance vision needs with a large share of low-light activities, such as night driving, it would be reasonable to recommend a monofocal lens or blended vision solution, while an advanced optics IOL would be preferable for the patient with a range of activities with the viewing distance distributed between all vision zones and requiring often switching between zones.

The information provided by the VBM allows healthcare professionals to objectively assess the needs of the patient and explain the benefits and limitations of the available solutions to the patient. However, the decision should include other factors, like personal preferences of the patient and tolerance to the multifocality of IOLs with advanced optics, as well as other conditions and, thus, should not exclude discussion with the patient. The measuring elements used in the VBM are standard components. Furthermore, the VBM is not intended to replace state-of-the art diagnostic methods in refractive treatment. Nevertheless, it is very important to have hard facts to support the choice of IOL suitability or surgical method.

Patients use the VBM in their normal daily routine, with the only excluded activities being those that involve water exposure of the wearable. Some other activities may be exempt based on a benefit/risk assessment. Those activities are meant to be communicated verbally to the attending HCP during pre-operative interview.

A dedicated tablet is provided to the clinics for measurement data upload on the cloud and data review. The embedded cellular data module of the tablet is used for data upload and report review on the cloud, independent of the clinic's infrastructure.

The clinic has access to a detailed report of all data the VBM collects. A simplified patient report including recommendations is available for the patient.

The healthcare professional has an option to review aggregated metrics of visual behavior of the patient. For example, the viewing distance from all measurement days, collected by the patient, can be displayed as a histogram of the additional optical power (refraction) of the eye required to accommodate the objects of visual activity (Figure 2).

Further, it is possible to relate the viewing distances to the particular head inclinations in order to visualize viewing distances in a two-dimensional plane (Figure 3). Such visualization is helpful to educate patients on the relation of activities and visual distances as visual zones can be linked to the actual physical space.

Additionally, the system allows us to review activities chronologically, as performed by the patient during the selected measurement day. The distribution of viewing distances can be overlaid with other relevant parameters, such as ambient light exposure (Figure 4).

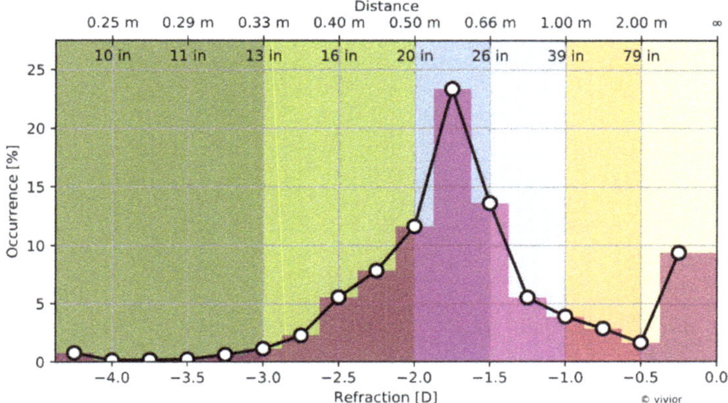

Figure 2. Distribution of viewing distance and required additional optical power of the eye (refraction) to observe objects of visual activity. The plot demonstrates domination of the intermediate viewing distance related to the computer work at around −1.75 D (0.57 m) for the particular patient (Patient A). Color coding is used to display the viewing zone, with green representing near viewing distances, blue representing intermediate, and yellow representing far.

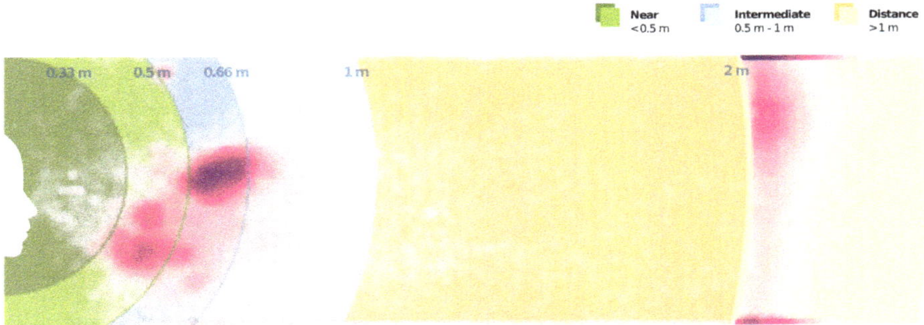

Figure 3. Mapping of viewing distances on two-dimensional plane using head flexion information (Patient A). Dense areas reflect the computer work and handheld devices. Color coding for the vision zones is the same as in Figure 2.

This kind of plot allows the demonstration of viewing conditions during various activities to the patient. The time of the day information helps the patient to relate his or her daily routine to the measurements and communicate to the HCP the importance of certain activities. The HCP is also able to relate viewing distance to other parameters. Figure 5 demonstrates the history of ultraviolet light exposure during the study day together with ambient light, which provides additional information about the patient's lifestyle.

The study has been conducted by five sites in three European countries: Germany, Switzerland, and Ireland (Table 1). A total of 129 patients were enrolled between August 2018 and May 2019. Patient age ranged from 49 to 82 (mean: 65). In all cases, the VBM measurement was performed before cataract surgery.

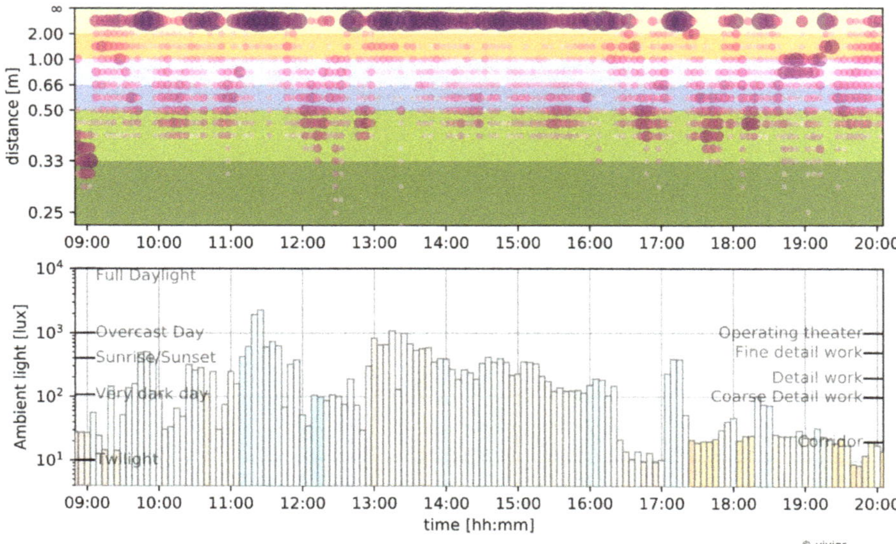

Figure 4. Example of the viewing distances and ambient light measurements with VBM during a day (Patient B). The example demonstrates the clear domination of far distances (upper plot), which is a marked difference in visual behavior from Patient A.

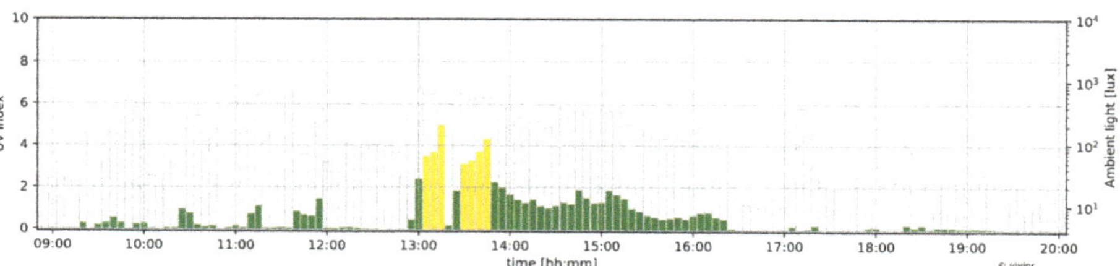

Figure 5. Example of the ultraviolet light exposure (quantified with UV index) as solid colored bars compared to ambient light level shown as empty bars during a measurement day shown in Figure 4 (Patient B). The level of ultraviolet light indicates that the example day was mostly spent outdoors.

Table 1. Vivior feasibility study sites, ethics committees, and enrollment.

Study Center	Patients Enrolled	Ethics Committee Approval
Wellington Eye Clinic, Dublin Ireland	35	Beacon Hospital Research Ethics Committee: BEA0096
University Eye Clinic Bochum-Langendreer, Bochum, Germany	40	Ruhr University Ethics Commission of Medical Faculty: CIV-18-05-023986
Sehkraft Augenzentrum Köln Germany	3	Ruhr University Ethics Commission of Medical Faculty: CIV-18-05-023986
Orasis Eye Clinic, Reinach, Switzerland	40	Swiss Ethics: Ethikkommission Nordwest- und Zentralschweiz: 2018-01344
LASER VISTA, Basel Switzerland	11	Swiss Ethics: Ethikkommission Nordwest- und Zentralschweiz: 2018-01344
Total	129	

All subjects gave their informed consent for inclusion before they participated in the study. The patients were informed in great detail about the content of the work, its usefulness, and its feasibility before the start of the study. The study was conducted in accordance with the Declaration of Helsinki, and the protocol was approved by the appropriate ethics committees for each site (Table 1).

3. Results

Of the total 178 eyes implanted, 119 eyes were implanted with monofocal intraocular lenses (67%), 39 eyes were implanted with trifocals or multifocals IOLs (22%), 15 eyes were implanted with toric lenses (8%), and the remaining five eyes were implanted with lenses that were not identified by the surgeon (3%). From these 129 patients included in the study, 178 cataract surgeries ultimately resulted, i.e., both eyes were not operated on in all cases, but only those that had an indication for surgery.

The pre-operative cataract scores distribution is shown in Figure 6 and the Cat-Quest 9 Task results are shown in Figure 7. A grade 3 or greater cataract manifested itself in 34% of the eyes, while no cataract and no reported cataract were accounted for in 29% of the eyes. The cataract and presbyopia limitations were also reflected in the pre-operative vision questionnaire with the vast majority of patients reporting some to very great difficulties in mastering everyday life and the majority of the patients being dissatisfied with the current eyesight.

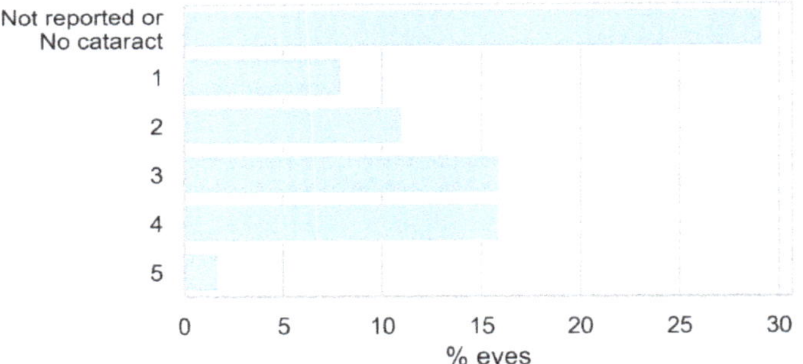

Figure 6. Lens opacities classification of cataract study patients.

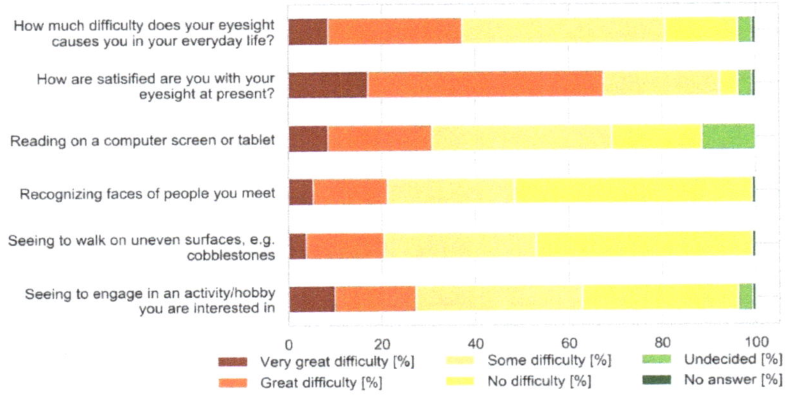

Figure 7. Pre-operative Cat-Quest 9 task results.

We have also investigated whether the difficulties in using the VMB system are related to the patients' visual problems. Based on the patients' answers to the usability questionnaire, the data were statistically evaluated using the parameters of corrected distant visual acuity and the Lens Opacities Classification System (LOCS) score. These results are shown in Table 2.

Table 2. Distribution of the preoperative corrected distance visual acuity and Lens Opacities Classification System (LOCS) III score of the patients separated by the answers in the usability questionnaire.

Question	Yes			No		
	N	LOCS III Score Mean (Std)	CDVA logMAR Mean (Std)	N	LOCS III Score Mean (Std)	CDVA logMAR Mean (Std)
1. Is it easy to attach the magnetic clip to your frames?	113	2.087 (1.179)	0.209 (0.215)	10	1.500 (1.323)	0.118 (0.161)
2. Is it easy to attach the wearable to the magnetic clip?	118	2.093 (1.178)	0.206 (0.213)	5	0.625 (0.750)	0.111 (0.197)
3. Do you always feel comfortable using the wearable?	98	2.203 (1.178)	0.233 (0.219)	25	1.333 (1.017)	0.080 (0.124)
4. Is it easy to use the wearable?	111	2.053 (1.186)	0.208 (0.217)	11	1.889 (1.364)	0.148 (0.161)
5. Is the connection between the station and wearable always working?	82	2.175 (1.178)	0.221 (0.209)	22	1.525 (1.282)	0.192 (0.220)
6. Is the information provided on the screen always clear to you?	71	1.885 (1.117)	0.172 (0.189)	44	2.188 (1.353)	0.236 (0.222)

Surprisingly, the groups that had difficulties with the VBM presented the higher visual acuity (lower logMAR values), as well as lower cataract scores (better vision), for all the questions, except for one concerning "information provided on the screen." The study included 15 patients with both eyes cataract-free and a relatively high distance-corrected visual acuity preparing for the clear lens exchange. This group was typically more demanding toward the usability of the system and often appeared in the group answering "no" in the questionnaire in Figure 8. This resulted in the proportion of patients without cataract in the "no" group being between 20% and 40%, while the mean was around 12%. The exception was question 6 where there was only a slightly higher proportion of clear lens patients in the "no" group (16% vs. 11%). Based on this data, we were not able to statistically attribute the difficulties reported in the usability questionnaire to the vision deficiencies of the patients for all questions except question 6, which also reported the highest negative feedback (34% not agreeing that the information provided on the screen is always clear).

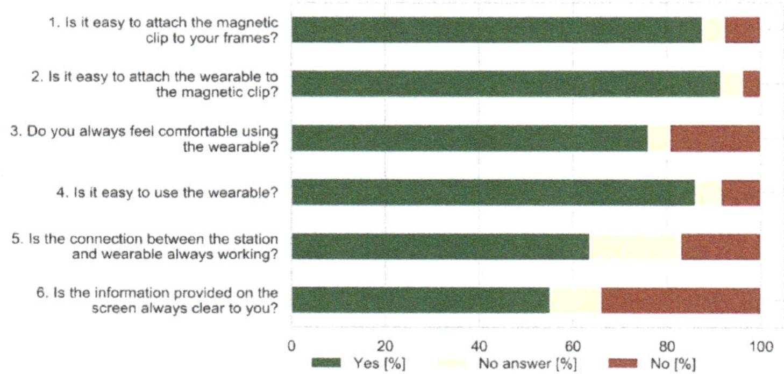

Figure 8. Results summary of usability questionnaire.

We have also studied the patient satisfaction with the vision results (Figure 9).

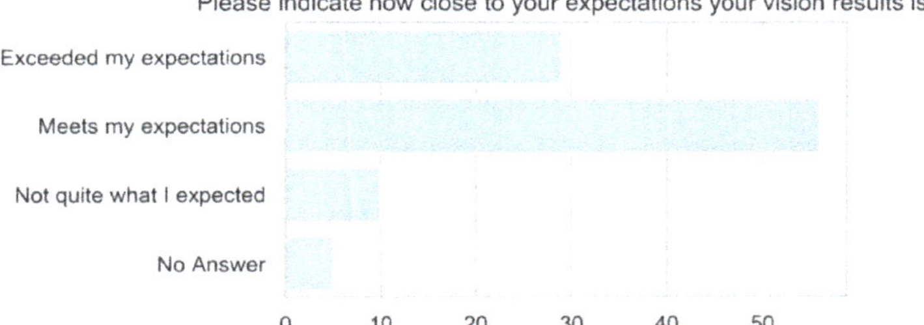

Figure 9. Post-operative Cat-Quest 9 results showing that surgical results met or exceeded patients' expectations.

A high level of patient satisfaction was reported, with 84% of patients meeting or even exceeding expectations (Figure 9).

4. Discussion

In today's world, visual acuity is not only becoming more and more important, but the demands on the visual system have increased significantly, whether at work or in leisure time. Various studies have shown that 61% of the patients prefer an explicit decision by the physician regarding the cataract surgery [21,22]. This strong dependency and importance of the physician's decision-making must be compared to the importance of patient involvement in order to maintain high patient satisfaction [23,24]. Previous studies show that the improving patients' perceived level of understanding is an important factor for the high level of patient satisfaction with the results of a cataract surgery [1,25–27]. There is an increasing demand to understand patient needs and expectations when planning cataract and refractive surgeries. It is generally accepted that understanding patient behavior and expectations and including this information in the planning process results in better surgical outcomes. Psychometric tests in the form of cataract surgery outcome questionnaires have been developed, identifying the limitations of performing daily tasks due to an individual's vision, a trait known as visual functioning [15,16]. The use of questionnaires and visual assessments in measuring patient outcomes is effective; however, the subjectivity of these methods is questionable. Nevertheless, questionnaires are very important in addition to objective measurement, because they reflect the important subjective part of the patient's feelings. Its importance and introduction have been used in the literature for a long time where specific questionnaires have been developed for specific questions [23], as we have used it in our study. Current psychometric measures may be subject to systematic error such as recall bias and could cause judgment errors resulting in patient discontent with surgical outcomes [19,20]. Integrating technology, such as the wearable VBM, which collects objective visual behavior data for pre-operative patient assessments, has shown improvement in the current understanding of patient visual behavior. It was very gratifying to note that the application of the VBM was, in the vast majority of cases, problem-free. Thus, the VBM measuring device could be installed with reasonable effort and be reliably used by the patient. In this sense, the goal was achieved to design a VBM in a way that not only works reliably but is also easy to use. Most of the study patients stated that they had severe to very severe visual problems before the surgery. None of the patients who reported very great difficulties in their daily tasks reported troubles in attaching the clip to the spectacles frame or VBM to the clip. The results of the analysis of the patients' visual performance based on their responses to the questionnaire showed that the group that was not satisfied with the user-friendliness had, on average, a higher visual acuity and a lower LOCS score, with the exception of one question related to the presentation of the information on the screen. This effect of the higher reported usability problems of patients with better vision can be attributed to the higher demands of patients with clear natural lenses. The feedback from patients on the system handling was mostly related to the adaptation to a specific spectacle frame, the correct alignment of the VBM and clip, and the strength of the magnetic attachment. This feedback has been considered in the development of the next generation of the system. The size of the clip has been adapted to make it easier to handle, and the shape of the clip and wearable has been redesigned to visually match each other to make the attachment more intuitive. The clip has a defined arrow shape that points forward to indicate the direction of attachment. The magnetic attachment has been replaced by a purely mechanical attachment to prevent the design from falling off during heavy movement, while at the same time being easy to attach and detach. The reported difficulties in understanding the information on the station's screen and connection problems between the wearable and the station have led to the decision to make the next unit self-sufficient and to omit an additional station for data acquisition, thus eliminating both station-related difficulties.

The objective data generated could aid clinicians in providing a highly individualized evaluation of a patient's visual needs, including estimated viewing distances, and determine the right course of action to improve post-operative patient outcomes and patient satisfaction. The subjective and objective data of patients often coincide. However, the patient's subjective feelings are not always correctly interpreted by them. The objective data collected can close this circle to a higher percentage and thus have potential to increase patient satisfaction. Including objective data of the visual behavior of each individual patient is another important step for a successful customized treatment.

Prior to cataract- and refractive surgery in addition to a comprehensive eye exam in terms of the overall health of the eyes, refraction measurements (amount of nearsightedness, farsightedness, astigmatism), determining the corneal curvature, as well as the length of the eye, it is particularly important to analyze the environmental conditions in the eye and ophthalmological concomitant diseases. The objectively measured ophthalmological findings, as well as the VBM data collected, have the potential to further improve treatment outcomes.

There is no ideal correction for all distances currently available, but the objective measurements of viewing distances, ambient light conditions, and head inclinations with devices such as the VBM will potentially allow better characterization of a patient's lifestyle and support the choice of IOL solutions better matching a specific patient's lifestyle and visual needs. Further studies with extended questions are ongoing.

5. Conclusions

To test the VBM device usability among patients, a multi-center feasibility study was conducted. Patients found the device easy to use, with most reporting that the device was not intrusive. The patients, even those that were elderly with significant cataracts, were able to comfortably use the VBM. Hence, the results of this feasibility study demonstrated that the concept of wearable monitoring of visual behavior is, in general, accepted by the patients. VBM may be used correctly and reliably by the patient. The difficulties of handling the system by the patients have been considered in the design of the next generation of the VBM.

The objective data collected with the VBM may have the potential to improve treatment outcomes, and the next research phase will seek to objectively assess the impact of the VBM on surgeons' treatment decision making.

Author Contributions: B.P. provided the treatment indication, developed the study design, acquired clinical data, and contributed to writing the paper. Z.C. contributed substantially to the methodology and substantially contributed to writing the paper. B.P.-E. performed substantial data curation. P.Z. performed data analysis and contributed to the study design. All authors have read and agreed to the published version of the manuscript.

Funding: This research received no external funding.

References

1. Pager, C.K. Expectations and outcomes in cataract surgery: A prospective test of 2 models of satisfaction. *Arch. Ophthalmol.* **2004**, *122*, 1788–1792. [CrossRef] [PubMed]
2. Wang, W.; Yan, W.; Fotis, K.; Prasad, N.M.; Lansingh, V.C.; Taylor, H.R.; Finger, R.P.; Facciolo, D.; He, M. Cataract Surgical Rate and Socioeconomics: A Global Study. *Investig. Ophthalmol. Vis. Sci.* **2016**, *57*, 5872–5881. [CrossRef] [PubMed]
3. Aristodemou, P.; Sparrow, J.M.; Kaye, S. Evaluating Refractive Outcomes after Cataract Surgery. *Ophthalmology* **2019**, *126*, 13–18. [CrossRef] [PubMed]
4. Pajic, B.; Massa, H.; Eskina, E. Presbyopiekorrektur mittels Lasechirurgie. *Klin. Monatsbl. Augenheilkd.* **2017**, *234*, e29–e42. [PubMed]
5. Knorz, M.C. Presbyopia Correction with Intraocular Lenses. *Klin. Monbl. Augenheilkd.* **2020**, *237*, 213–223. [PubMed]
6. Goldberg, D.G.; Goldberg, M.H.; Shah, R.; Meagher, J.N.; Ailani, H. Pseudophakic mini-monovision: High patient satisfaction, reduced spectacle dependence, and low cost. *BMC Ophthalmol.* **2018**, *18*, 293. [CrossRef] [PubMed]
7. Tarib, I.; Kasier, I.; Herbers, C.; Hagen, P.; Breyer, D.; Kaymak, H.; Klabe, K.; Lucchesi, R.; Teisch, S.; Diakonis, V.F.; et al. Comparison of Visual Outcomes and Patient Satisfaction after Bilateral Implantation of an EDOF IOL and a Mix-and-Match Approach. *J. Refract Surg.* **2019**, *35*, 408–416. [CrossRef]
8. Hovanesian, J.A.; Lane, S.S.; Allen, Q.B.; Jones, M. Patient-Reported Outcomes/Satisfaction and Spectacle Independence with Blended or Bilateral Multifocal Intraocular Lenses in Cataract Surgery. *Clin. Ophthalmol.* **2019**, *13*, 2591–2598. [CrossRef]
9. Shih, W.; Liu, J. A calibration-free gaze tracking technique. In Proceedings of the 15th Conference Patterns Recognition, Barcelona, Spain, 3–7 September 2000; Volume 4, pp. 201–204.
10. Ji, Q.; Yang, X. Real-time eye, gaze and face pose tracking for monitoring driver vigilance. *Real-Time Imaging* **2002**, *8*, 357–377. [CrossRef]
11. Grace, R. Drowsy driver monitor and warning system. In Proceedings of the International Driving Symposium on Human Factors Driver Assessment, Training and Vehicle Design, Aspen, CO, USA, 14–17 August 2001; pp. 64–69.
12. Bergasa, L.M.; Nuevo, J.; Sotelo, M.A.; Barea, R.; Lopez, M.E. Real-Time System for Monitoring Driver Vigilance. In *IEEE Transactions on Intelligent Transportation Systems*; IEEE: New York, NY, USA, 2006; Volume 7.
13. Lodhia, V.; Karanja, S.; Lees, S.; Bastawrous, A. Acceptability, usability, and views on deployment of peek, a mobile phone mHealth intervention for eye care in Kenya: Qualitative study. *JMIR mHealth uHealth* **2016**, *4*, e30. [CrossRef]
14. Groessl, E.J.; Liu, L.; Sklar, M.; Tally, S.R.; Kaplan, R.M.; Ganiats, T.G. Measuring the impact of cataract surgery on generic and vision-specific quality of life. *Qual. Life Res.* **2013**, *22*, 1405–1414. [CrossRef] [PubMed]
15. Stelmack, J. Quality of life of low vision patients and outcomes of low vision rehabilitation. *Optom. Vis. Sci.* **2001**, *78*, 335–342. [CrossRef] [PubMed]
16. McAlinden, C.; Gothwal, V.K.; Khadka, J.; Wright, T.A.; Lamoureux, E.L.; Pesudovs, K. A head-to-head comparison of 16 cataract surgery outcome questionnaires. *Ophthalmology* **2001**, *118*, 2374–2381. [CrossRef] [PubMed]
17. Lundström, M.; Pesudovs, K. Catquest-9SF patient outcomes questionnaire. Nine-item short-form Rasch-scaled revision of the Catquest questionnaire. *J. Cataract Refract. Surg.* **2009**, *35*, 504–513. [CrossRef] [PubMed]
18. Morlock, R.; Wirth, R.J.; Tally, S.R.; Garufis, C.; Heichel, C.W.D. Patient-Reported Spectacle Independence Questionnaire (PRSIQ): Development and Validation. *Am. J. Ophthalmol.* **2017**, *178*, 101–113. [CrossRef]
19. Nogueira, H.N.; Alves, M.; Schor, P. Preoperative automatic visual behavioural analysis as a tool for intraocular lens choice in cataract surgery. *Arq. Bras. Oftalmol.* **2015**, *78*, 94–99. [CrossRef]
20. Gibbons, A.; Ali, T.K.; Waren, D.P.; Donaldson, K.E. Causes and correction of dissatisfaction after implantation of presbyopia-correcting intraocular lenses. *Clin. Ophthalmol.* **2016**, *10*, 1965–1970. [CrossRef]
21. Kiss, C.G.; Richter-Mueksch, S.; Stifter, E.; Diendorfer-Radner, G.; Velikay-Parel, M.; Radner, W. Informed consent and decision making by cataract patients. *Arch. Ophthalmol.* **2004**, *122*, 94–98. [CrossRef]

22. Henderson, B.A.; Solomon, K.; Masket, S.; Potvin, R.; Holland, E.J.; Cionni, R.; Sandoval, H. A survey of potential and previous cataract-surgery patients: What the ophthalmologist should know. *Clin. Ophthalmol.* **2014**, *25*, 1595–1602.

23. Berg, A.; Yuval, D.; Ivancovsky, M.; Zalcberg, S.; Dubani, A.; Benbassat, J. Patient perception of involvement in medical care during labor and delivery. *Israel Med. Assoc. J.* **2001**, *3*, 352–356.

24. Janz, N.K.; Wren, P.A.; Copeland, L.A.; Lowery, J.C.; Goldfarb, S.L.; Wilkins, E.G. Patient-physician concordance: Preferences, perceptions, and factors influencing the breast cancer surgical decision. *J. Clin. Oncol.* **2004**, *1*, 3091–3098. [CrossRef] [PubMed]

25. Addisu, Z.; Solomon, B. Patients' preoperative expectation and outcome of cataract surgery at jimma university specialized hospital-department of ophthalmology. *Ethiop. J. Health Sci.* **2011**, *21*, 47–55. [CrossRef] [PubMed]

26. Chen, Z.; Lin, X.; Qu, B.; Gao, W.; Zuo, Y.; Peng, W.; Jin, L.; Yu, M.; Lamoureux, E. Preoperative Expectations and Postoperative Outcomes of Visual Functioning among Cataract Patients in Urban Southern China. *PLoS ONE* **2017**, *9*, e0169844. [CrossRef] [PubMed]

27. Leidy, N.K.; Beusterien, K.; Sullivan, E.; Richner, R.; Muni, N.I. Integrating the patient's perspective into device evaluation trials. *Value Health* **2006**, *9*, 394–401. [CrossRef] [PubMed]

Article

Pressure Observer Based Adaptive Dynamic Surface Control of Pneumatic Actuator with Long Transmission Lines

Deyuan Meng [1,*], Bo Lu [2], Aimin Li [1], Jiang Yin [1] and Qingyang Li [1]

[1]   School of Mechatronic Engineering, China University of Mining and Technology, Xuzhou 221116, China
[2]   National Quality Supervision and Inspection Centre of Pneumatic Products, Ningbo 315500, China
*   Correspondence: tinydreams@126.com; Tel.: +86-1516-2126-417

Received: 22 July 2019; Accepted: 26 August 2019; Published: 3 September 2019

Featured Application: This research dedicates to develop a pressure observer based nonlinear controller for precise position control of the MRI-compatible pneumatic servo system with long transmission lines.

Abstract: In this paper, the needle insertion motion control of a magnetic resonance imaging (MRI) compatible robot, which is actuated by a pneumatic cylinder with long transmission lines, is considered and a pressure observer based adaptive dynamic surface controller is proposed. The long transmission line is assumed to be an intermediate chamber connected between the control valve and the actuator in series, and a nonlinear first order system model is constructed to characterize the pressure losses and time delay brought by it. Due to the fact that MRI-compatible pressure sensors are not commercially available, a globally stable pressure observer is employed to estimate the chamber pressure. Based on the model of the long transmission line and the pressure observer, an adaptive dynamic surface controller is further designed by using the dynamic surface control technique. Compared to the traditional backstepping design method, the proposed controller can avoid the problem of "explosion of complexity" since the repeated differentiation of virtual controls is no longer required. The stability of the closed-loop system is analytically proven by employing the Lyapunov theory. Extensive experimental results are presented to demonstrate the effectiveness and the performance robustness of the proposed controller.

Keywords: pneumatic servo system; long transmission line; pressure observer; dynamic surface control; position tracking

---

1. Introduction

The magnetic resonance imaging (MRI) technique is widely used in clinical diagnosis due to its ability to image without the use of ionizing x-rays and superior soft tissue contrast as compared to computed tomography (CT) scanning. Recently, pneumatically actuated MRI-compatible robots, which enable real-time magnetic resonance (MR) image-guided needle placement, are designed for brachytherapy and biopsy by several researchers [1–15]. To fulfill the requirements for MR compatibility of the robotic systems, pneumatic valves are commonly placed outside the scanner room in the aforementioned works. Therefore, long transmission lines between the actuators and valves are used. Since long transmission lines have a significant influence on the pressure dynamics of the pneumatic system and MRI-compatible pressure sensors are not commercially available, precise position control is one of the main technical challenges in robot development.

Indeed, considerable research effort has been devoted to addressing the issue related to long transmission lines. Richer et al. [16] suggested a formula for the time delay and amplitude attenuation between the mass flows at the outlet and inlet of the line, as well as a formula for the pressure drop along the tube. Although this approach was later employed in many other studies (Jiang et al. [14]; Richer et al. [17]), it has been proven improper for long transmission lines (>2 m) [18]. Yang et al. [8] adopted a first order transfer function with time delay to describe the 9-m transmission line dynamics. However, for some unknown reason, the authors omitted the transmission line dynamics in the subsequent controller design. Based on the discretization of a general set of equations from fluid mechanics, two similar distributed models of long pneumatic transmission lines were derived by Li et al. [19] and Krichel et al. [20]. However, this type of model is not suitable for controller design for their high order and low computational speed. Turkseven et al. [21] developed a simplified distributed model to characterize the additional dynamics brought by the long transmission lines.

As previously mentioned, direct pressure measurement is unavailable for the lack of MRI-compatible pressure sensors. However, pressure states are commonly used in controller design for precise position control of a pneumatic actuator. To solve this problem, pressure observers were used in place of a pressure sensor in many studies. Pandian et al. [22] investigated two design methods of an observer to estimate the chamber pressures in a pneumatic cylinder. While a continuous gain observer was used to estimate one of the chamber pressures with the assumption that the other one is measured in the first method, a sliding mode observer was utilized to estimate both chamber pressures in the second method. In this study, the mass flow rate was assumed to be exactly known, which is apparently too restrictive in practice. Wu et al. [23] conducted an analysis on observability and concluded that the pressure states in the pneumatic servo system are not locally observable from the measurement of the output motion within several regions in the state space. Based on the actuator pressure dynamics, various observers are developed, such as the sliding-mode observer (Bigras [24]), energy-based Lyapunov observer (Gulati et al. [25]), and adaptive nonlinear observer (Langjord et al. [26]). It should be noted that the performance of these observers relies on the accuracy of the valve model. Driver et al. [27] developed a pressure estimation algorithm by utilizing the measured actuation force and the hypothesized average air pressure in the actuator. Turkseven et al. [18] presented a pressure observation method by using the measured force and piston displacement. However, the requirement for force sensing in the MRI environment is particularly burdensome because the MRI-compatible force sensor is not available commercially and its development cost is high.

Recently, the backstepping design method has been proven to be a very effective way to develop nonlinear robust controllers for pneumatic servo systems. However, this method has the problem of "explosion of complexity" since the requirement of repeated differentiation of virtual controls. Thus, a practical implementation is difficult. To solve this problem, Swaroop et al. [28] proposed the dynamic surface control (DSC) method, in which the calculation of the virtual control variable's derivative was prevented by introducing a filter at each design step. Since then, the DSC method has been the topic of significant research efforts and a number of excellent theoretical contributions have been made. The applications of DSC can be found in many engineering fields, for example, hydraulic servo systems [29], underwater/autonomous surface vehicles [30], mobile wheeled inverted pendulum [31], pneumatic artificial muscle [32], and servo motor [33].

In this study, the needle insertion motion control of the MRI compatible robot developed in our lab is considered. The robot is actuated by a pneumatic cylinder with long transmission lines. The focus of this paper is dealing with the issue of long transmission lines and realizing a high accuracy control of a pneumatic actuator with a pressure observer. Therefore, the long transmission line is assumed to be an intermediate chamber connected between the control valve and the actuator in series, and a pressure observer based adaptive dynamic surface control is proposed. The main contributions are: (1) The long transmission line dynamic is approximated as a nonlinear first order system, which can estimate the pressure losses and time delay through the long transmission line precisely in real time. (2) A pressure observer, which is proven to be globally stable, is developed to estimate the chamber pressure in a pneumatic actuator with a long transmission line. (3) In contrast to most of the existing nonlinear robust controllers synthesized by the backstepping method, by using the dynamic surface control

(DSC) technique, the proposed controller can cope with the problem of "explosion of complexity", since the repeated differentiation of virtual controls is no longer required. The rest of this paper is organized as follows: Section 2 presents the dynamic models and problem statement; Section 3 gives the design and stability proof of the pressure observer based adaptive dynamic surface controller; Section 4 presents the experimental results to demonstrate the performance of the proposed controller; and Section 5 draws the conclusions.

2. Dynamic Models and Problem Formulation

As shown in Figure 1, a 5-DOF (degree of freedom) MRI compatible robot for abdominal and thoracic punctures was built in our laboratory. The robot's mechanism design was developed such that all motions were decoupled and actuated by pneumatic cylinders. With the help of the robot, the physician could manipulate the needle remotely without moving the patient out of the MRI scanners. A schematic of the pneumatic servo system driving the needle insertion motion is depicted in Figure 2. The pneumatic cylinder (customized product from XMC Corp., Ningbo, China), which had a 150 mm stroke and 10 mm diameter bore, was made of nonmagnetic material to ensure MRI compatibility. While Chamber A was controlled by a proportional directional control valve (FESTO MPYE-5-M5-010-B), a tank was used to maintain the pressure of Chamber B at a specified constant level for safety reasons. The control valve was connected to the cylinder via a 10 m transmission line since it had to be placed away from the MRI scanner. A pressure observer was needed to estimate the pressure of Chamber A, since direct measurement was expensive for the lack of MRI-compatible pressure sensors. The main purpose of this paper was to find a way to deal with the issue of the long transmission line and realize high accuracy control of a pneumatic cylinder with pressure observation. For the purpose of comparison, the pressures at the two ends of the long transmission line, as well as the supply pressure were measured by three pressure sensors (FESTO SDET-22T-D10-G14-I-M12). The piston position was obtained with an optical position sensor (Micronor MR328). The controller was programmed in Simulink on a dSPACE DS1103 control system with a 1 ms sampling period.

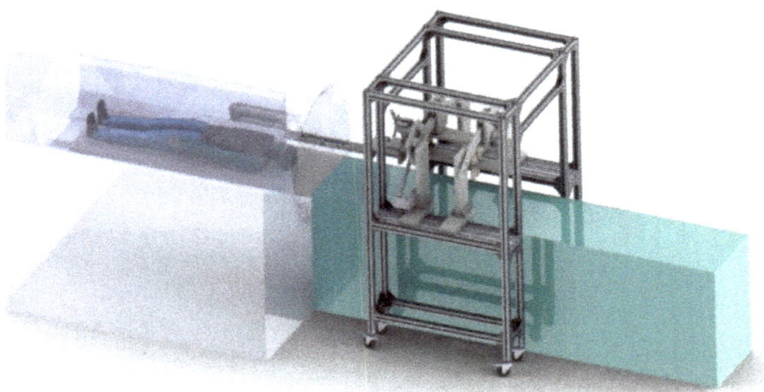

Figure 1. Virtual prototype of a 5-DOF (degree of freedom) magnetic resonance imaging (MRI) compatible robot for abdominal and thoracic punctures.

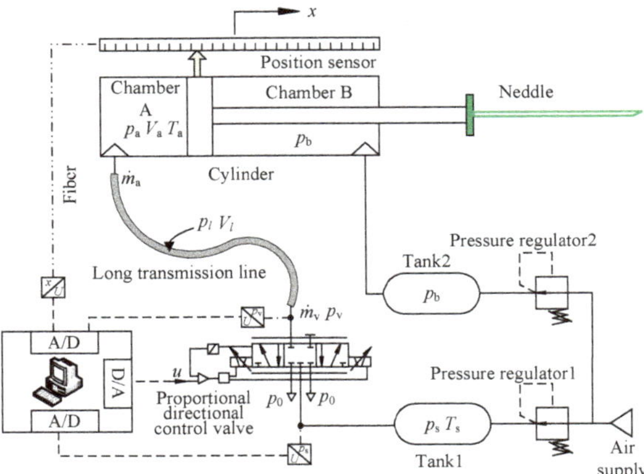

Figure 2. Schematic of the pneumatic servo system used to drive the needle insertion motion.

As shown in [34], the motion of the piston–rod–needle assembly can be expressed as:

$$m\ddot{x} = p_a A_a - p_b A_b - p_0 A_r - b\dot{x} - F_f S_f(\dot{x}) - d_f, \tag{1}$$

where x, $\dot{x}$, and $\ddot{x}$ are the piston position, velocity, and acceleration, respectively, m is the normal mass of the piston–rod–needle assembly, p_a and p_b denotes the absolute pressures of actuator chambers, p_0 is the ambience pressure, A_a and A_b are the cross section areas of piston chambers, A_r is the cross section area of the rod, b is the total load and cylinder viscous friction coefficient, F_f is the unknown friction coefficient, $S_f(\dot{x})$ is a continuous function, which is always chosen as $S_f(\dot{x}) = \frac{2}{\pi}\arctan(900\dot{x})$, $F_f S_f(\dot{x})$ is utilized as the smooth approximation for the usual static discontinuous Coulomb friction force, and d_f represents the unmodeled dynamics and external disturbances.

With the assumption that the discharging and charging processes are both isothermal, the pressure dynamic in the actuator chamber can be modeled as [16,35]:

$$\dot{p}_a = \frac{RT_s}{V_a}\dot{m}_a - \frac{A_a p_a}{V_a}\dot{x}, \quad V_a = V_{a0} + A_a\left(\frac{L_c}{2} + x\right), \tag{2}$$

where R is the gas constant, T_s is the ambient temperature, V_a is the volume of the cylinder chamber A, $\dot{m}_a$ is the mass flow entering or exiting the chamber A, V_{a0} is the dead volume of the cylinder chamber A, and L_c is stroke of the actuator. $\dot{m}_a$ can be calculated by [16]:

$$\dot{m}_a = (p_l - p_a)\frac{p_v A_l D_l}{32 R T_s L_l \mu}, \tag{3}$$

where p_l is the average air pressure in the transmission line, p_v is the measured air pressure at the work port of the control valve, L_l, A_l, and D_l denote the length, cross section area, and inner diameter of the transmission line, respectively, and μ is the dynamic viscosity of air.

As shown in Figure 2, the long transmission line is assumed to be an intermediate chamber connected between the control valve and the actuator in series, thus, the following equation is formulated to represent the pressure dynamic of the line [16,35]:

$$\dot{p}_l = \frac{RT_s}{V_l}(\dot{m}_v - \dot{m}_a), \tag{4}$$

where V_l is the volume of the transmission line, and $\dot{m}_v$ is the mass flow rate through the valve, which can be calculated by [35]:

$$\dot{m}_v = A(u)K_q(p_u, p_d, T_u) = \begin{cases} A(u)C_dC_1\frac{p_u}{\sqrt{T_u}}, & \frac{p_d}{p_u} \leq p_r, \\ A(u)C_dC_1\frac{p_u}{\sqrt{T_u}}\sqrt{1-\left(\frac{\frac{p_d}{p_u}-p_r}{1-p_r}\right)^2}, & \frac{p_d}{p_u} > p_r \end{cases} \tag{5}$$

where $A(u)$ is the effective valve orifice area, u is the control valve's control input, C_d is the discharge coefficient, C_1 is a constant with a value of 0.0404, p_u and p_d are the upstream pressure and the downstream pressure, respectively, T_u is the upstream temperature of air, and p_r is the critical pressure ratio. Therefore, $\dot{m}_v = A(u)K_q(p_s, p_v, T_s)$ when the line is charging, and $\dot{m}_v = A(u)K_q(p_v, p_0, T_s)$ when the line is discharging.

Choose the state vectors $x = [x_1, x_2, x_3, x_4]^T$ as $x_1 = x$, $x_2 = \dot{x}$, $x_3 = p_a$ and $x_4 = p_l$, thus the system dynamics can be expressed in a state-space form as:

$$\begin{cases} \dot{x}_1 = x_2 \\ m\dot{x}_2 = A_a x_3 - A_b p_b - p_0 A_r - bx_2 - A_f S_f(x_2) - d_f \\ \dot{x}_3 = \frac{RT_s}{V_a}\dot{m}_a - \frac{A_a x_2}{V_a}x_3 \\ \dot{x}_4 = \frac{RT_s}{V_l}(\dot{m}_v - \dot{m}_a) \end{cases} \tag{6}$$

The control objective is to synthesize a control input u for the system Equation (6) such that x tracks the desired trajectory x_d with a guaranteed transient and final tracking accuracy. Instead of using the pressure sensor, a globally stable pressure observer is developed to estimate the pressure in chamber A. The parametric uncertainties due to unknown b and A_f and unknown external disturbances will be explicitly considered in this paper.

Assumption 1: *The desired trajectory x_d is at least second-order differentiable, and x_d, $\dot{x}_d$, and $\ddot{x}_d$ are bounded. Then, there exists a compact set $\Omega_0 = \left\{\left[x_d, \dot{x}_d, \ddot{x}_d\right]^T : x_d^2 + \dot{x}_d^2 + \ddot{x}_d^2 \leq B_0\right\}$ such that $\left[x_d, \dot{x}_d, \ddot{x}_d\right]^T \in \Omega_0$, where B_0 is a positive constant.*

Assumption 2: *The extent of parametric uncertainties and external disturbances can be predicted and given by $b_{min} \leq b \leq b_{max}$, $F_{fmin} \leq F_f \leq F_{fmax}$, and $d_{min} \leq d_f \leq d_{max}$.*

3. Controller Design

3.1. Pressure Observer

The following observer is proposed to estimate the pressure of the actuator chamber A and the average air pressure in the transmission line:

$$\begin{cases} \dot{\hat{p}}_a = \frac{RT_s}{V_a}\hat{\dot{m}}_a - \frac{A_a\dot{x}}{V_a}\hat{p}_a \\ \dot{\hat{p}}_l = \frac{RT_s}{V_l}(\hat{\dot{m}}_v - \hat{\dot{m}}_a) \end{cases} \tag{7}$$

where $\hat{p}_a$ and $\hat{p}_l$ represent the estimates of p_a and p_l, $\hat{\dot{m}}_a$ and $\hat{\dot{m}}_v$ denote the estimated mass flow rates according to Equation (3) and Equation (5) based on $\hat{p}_a$ and $\hat{p}_l$. It should be noted that the proposed observers are closed loop because of the relationship between the estimated mass flow rates and the estimated pressures. Proof of the convergence between the actual pressures and the estimated ones will be given later.

To verify the convergence between the actual pressures and the estimated pressures, the following Lyapunov function candidate was considered:

$$V_o = V_{o1} + V_{o2}, V_{o1} = \frac{1}{2}(\widetilde{p}_a V_a)^2, V_{o2} = \frac{1}{2}(\widetilde{p}_l V_l)^2, \tag{8}$$

where $\widetilde{p}_a = p_a - \hat{p}_a$, and $\widetilde{p}_l = p_l - \hat{p}_l$.

The time derivative of V_{o1} is:

$$\begin{aligned}
\dot{V}_{o1} &= (p_a V_a - \hat{p}_a V_a)(\dot{p}_a V_a + p_a \dot{V}_a - \dot{\hat{p}}_a V_a - \hat{p}_a \dot{V}_a) \\
&= RT_s V_a (p_a - \hat{p}_a)(\dot{m}_a - \hat{\dot{m}}_a)
\end{aligned} \tag{9}$$

According the Equation (3), one can obtain:

1. $p_a > \hat{p}_a$ implies that $\dot{m}_a < \hat{\dot{m}}_a$, yielding $(p_a - \hat{p}_a)(\dot{m}_a - \hat{\dot{m}}_a) < 0$.
2. $p_a < \hat{p}_a$ implies that $\dot{m}_a > \hat{\dot{m}}_a$, yielding $(p_a - \hat{p}_a)(\dot{m}_a - \hat{\dot{m}}_a) < 0$.
3. $p_a = \hat{p}_a$ implies that $\dot{m}_a < \hat{\dot{m}}_a$, yielding $(p_a - \hat{p}_a)(\dot{m}_a - \hat{\dot{m}}_a) = 0$.

Therefore, $\dot{V}_{o1}$ is negative semi-definite.

The time derivative of V_{o2} is:

$$\begin{aligned}
\dot{V}_{o2} &= (p_l V_l - \hat{p}_l V_l)(\dot{p}_l V_l - \dot{\hat{p}}_l V_l) \\
&= RT_s V_l (p_l - \hat{p}_l)(\dot{m}_v - \hat{\dot{m}}_v) - RT_s V_l (p_l - \hat{p}_l)(\dot{m}_a - \hat{\dot{m}}_a)
\end{aligned} \tag{10}$$

According the Equation (3), one can obtain:

1. $p_l > \hat{p}_l$ implies that $\dot{m}_a > \hat{\dot{m}}_a$, yielding $(p_l - \hat{p}_l)(\dot{m}_a - \hat{\dot{m}}_a) > 0$.
2. $p_l < \hat{p}_l$ implies that $\dot{m}_a < \hat{\dot{m}}_a$, yielding $(p_l - \hat{p}_l)(\dot{m}_a - \hat{\dot{m}}_a) > 0$.
3. $p_l = \hat{p}_l$ implies that $\dot{m}_a = \hat{\dot{m}}_a$, yielding $(p_l - \hat{p}_l)(\dot{m}_a - \hat{\dot{m}}_a) = 0$.

For the mass flow rate relationship given by Equation (5), it has proven in [25] that the term $(p_l - \hat{p}_l)(\dot{m}_v - \hat{\dot{m}}_v)$ is always non-positive. Thus, $\dot{V}_{o2}$ and $\dot{V}_o$ are negative semi-definite, one can conclude that the proposed pressure observers are globally Lyapunov stable with regard to the errors in the estimated pressures.

3.2. Adaptive Dynamic Surface Controller

Step 1: Differentiate the trajectory tracking error $e_1 = x_1 - x_d$ with respect to time leads to:

$$\dot{e}_1 = x_2 - \dot{x}_d. \tag{11}$$

Consider x_2 as the first virtual control input, the virtual control law is designed as follows:

$$\bar{x}_{2d} = \dot{x}_d - k_1 e_1, \qquad k_1 > 0, \tag{12}$$

where k_1 is a positive control gain. Following the dynamic surface control theory [28], the signal $\bar{x}_{2d}$ is fed to a low pass filter to obtain a new one x_{2d} for the next design step:

$$\tau_1 \dot{x}_{2d} + x_{2d} = \bar{x}_{2d}, \qquad x_{2d}(0) = \bar{x}_{2d}(0), \qquad \tau_1 > 0, \tag{13}$$

where τ_1 is the filter parameter.

Step 2: Define the first surface error as $s_1 = x_2 - x_{2d}$, its time derivative can be derived as:

$$\dot{s}_1 = \dot{x}_2 - \dot{x}_{2d} = \bar{A}_a x_3 - \frac{1}{m}\left[p_b A_b + p_0 A_r + b x_2 + F_f S_f(\dot{x}) + d_f\right] - \frac{1}{\tau_1}(\bar{x}_{2d} - x_{2d}), \tag{14}$$

where $\overline{A}_a = A_a/m$.

Define $\widetilde{x}_3 = x_3 - \hat{x}_3$, and let $\varsigma = \hat{x}_3$ be the second virtual control input, the virtual control law is designed as follows:

$$\overline{\varsigma}_d = \frac{1}{\overline{A}_a}\left\{\frac{1}{m}\left[p_bA_b + p_0A_r + \hat{b}x_2 + \hat{F}_fS_f(\dot{x}) + \hat{d}_f\right] + \frac{1}{\tau_1}(\overline{x}_{2d} - x_{2d}) - k_2s_1 - \frac{h_1^2}{4\eta_1}s_1\right\}, \tag{15}$$

where $k_2 > 0$ is a positive control gain, $\eta_1 > 0$ is a controller parameter, and h_1 is a known function that will be determined later, $\hat{b}$, $\hat{F}_f$, and $\hat{d}_f$ are the estimates of b, F_f, and d_f, respectively. $\hat{b}$, $\hat{F}_f$, and $\hat{d}_f$ are updated by:

$$\dot{\hat{b}} = \mathrm{Proj}_{\hat{b}}\left(-\frac{\gamma_1}{m}x_2s_1\right), \tag{16}$$

$$\dot{\hat{F}}_f = \mathrm{Proj}_{\hat{F}_f}\left(-\frac{\gamma_2}{m}S_f(x_2)s_1\right), \tag{17}$$

$$\dot{\hat{d}}_f = \mathrm{Proj}_{\hat{d}_f}\left(-\frac{\gamma_3}{m}s_1\right), \tag{18}$$

where $\gamma_1 > 0$, $\gamma_2 > 0$, and $\gamma_3 > 0$ are the observer gains, and the projection mapping is defined as:

$$\mathrm{Proj}_{\hat{\xi}}(\cdot) = \begin{cases} 0, & \text{if} \quad \hat{\xi} = \xi_{\max} \quad \text{and} \quad \cdot > 0 \\ 0, & \text{if} \quad \hat{\xi} = \xi_{\min} \quad \text{and} \quad \cdot < 0 \ , \\ \cdot, & \text{otherwise} \end{cases} \tag{19}$$

where ξ is a symbol that can be replaced by b, F_f, and d_f.

Similarly, the signal $\overline{\varsigma}_d$ is fed to a low pass filter to obtain a new one ς_d for the next design step:

$$\tau_2\dot{\varsigma}_d + \varsigma_d = \overline{\varsigma}_d, \varsigma_d(0) = \overline{\varsigma}_d(0), \ \tau_2 > 0, \tag{20}$$

where τ_2 is the filter parameter.

Step 3: Define the second surface error as $s_2 = \varsigma - \varsigma_d$, its time derivative can be derived as:

$$\begin{aligned}
\dot{s}_2 &= \dot{\varsigma} - \dot{\varsigma}_d = \dot{\hat{x}}_3 - \dot{\varsigma}_d = \frac{RT_s}{V_a}\hat{m}_a - \frac{A_ax_2}{V_a}\hat{x}_3 - \frac{1}{\tau_2}(\overline{\varsigma}_d - \varsigma_d) \\
&= -\frac{RT_sa}{V_a}\hat{x}_4 + \left(\frac{RT_sa}{V_a} - \frac{A_ax_2}{V_a}\right)\hat{x}_3 - \frac{1}{\tau_2}(\overline{\varsigma}_d - \varsigma_d)
\end{aligned} \tag{21}$$

where $a = \frac{p_vA_lD_l}{32RT_sL_l\mu}$.

Let $\zeta = -\hat{x}_4$ be the third virtual control input, the virtual control law is designed as follows:

$$\overline{\zeta}_d = \frac{V_a}{RT_sa}\left[\left(\frac{A_ax_2}{V_a} - \frac{RT_sa}{V_a}\right)\hat{x}_3 + \frac{1}{\tau_2}(\overline{\varsigma}_d - \varsigma_d) - k_3s_2\right], \tag{22}$$

where $k_3 > 0$ is a positive control gain. Similarly, the signal $\overline{\zeta}_d$ is fed to a low pass filter to obtain a new one ζ_d for the next design step:

$$\tau_3\dot{\zeta}_d + \zeta_d = \overline{\zeta}_d, \quad \zeta_d(0) = \overline{\zeta}_d(0), \ \tau_3 > 0, \tag{23}$$

where τ_3 is the filter parameter.

Step 4: Define the second surface error as $s_3 = \zeta - \zeta_d$, its time derivative can be derived as:

$$\dot{s}_3 = \dot{\zeta} - \dot{\zeta}_d = -\dot{\hat{x}}_4 - \dot{\zeta}_d = -\frac{RT_s}{V_l}(\hat{m}_v - \hat{m}_a) - \frac{1}{\tau_3}(\overline{\zeta}_d - \zeta_d). \tag{24}$$

Similarly, the following control law q_d for $\hat{m}_v$ is proposed:

$$q_d = \hat{m}_a - \frac{RT_s}{V_l}\left[\frac{1}{\tau_3}(\bar{\zeta}_d - \zeta_d) - k_4 s_3\right],\tag{25}$$

where $k_4 > 0$ is a positive control gain.

Once the q_d is calculated, the desired effective valve orifice area $A(u)$ can be calculated by:

$$A(u) = \begin{cases} \dfrac{q_d}{K_q(p_s,p_v,T_s)} & q_d > 0 \\[2mm] \dfrac{q_d}{K_q(p_v,p_0,T_s)} & q_d \leq 0 \end{cases}.\tag{26}$$

Thus, the input signal u for the proportional-directional control valve could be obtained according to the relation between the input signal and effective valve orifice area.

The proposed pressure observer based adaptive dynamic surface control is illustrated in Figure 3.

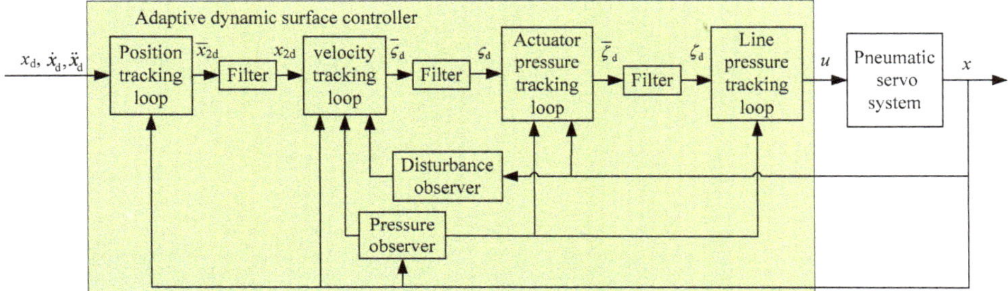

Figure 3. Block diagram of the control system.

3.3. Proof of Stability

Theorem 1. *Consider the closed loop pneumatic servo system consisting of the plant Equation (6), the nonlinear control law Equation (25), the pressure observers Equation (7), and the parameter and disturbance estimation algorithm Equations (16)–(18) under the Assumption 1–2. If there exists a set of the feedback gains and the filter constants satisfying* $\gamma = \min\left\{k_1 - 1, k_2 - \frac{1}{2} - \bar{A}_a, k_3 - \frac{\bar{A}_a}{2}, \frac{1}{\tau_1} - 1, \frac{1}{\tau_2} - \frac{1}{2} - \frac{\bar{A}_a}{2}\right\} > 0$, *then the closed loop system is uniformly and ultimately bounded.*

Proof. Consider the following Lyapunov function candidate:

$$\begin{aligned} V_c &= V_{cs} + V_{cz} + V_\theta, \quad V_{cs} = \tfrac{1}{2}e_1^2 + \tfrac{1}{2}s_1^2 + \tfrac{1}{2}s_2^2 + \tfrac{1}{2}s_3^2, \quad V_{cz} = \tfrac{1}{2}z_1^2 + \tfrac{1}{2}z_2^2 + \tfrac{1}{2}z_3^2, \\ V_\theta &= \tfrac{1}{2}\gamma_1^{-1}\tilde{b}^2 + \tfrac{1}{2}\gamma_2^{-1}\tilde{F}_f^2 + \tfrac{1}{2}\gamma_3^{-1}\tilde{d}_f^2 \end{aligned}\tag{27}$$

where $z_1 = x_{2d} - \bar{x}_{2d}, z_2 = \varsigma_d - \bar{\varsigma}_d, z_3 = \zeta_d - \bar{\zeta}_d$. Substituting Equation (12) into Equation (11) gives:

$$\dot{e}_1 = s_1 + z_1 - k_1 e_1.\tag{28}$$

Substituting Equation (15) into Equation (14) and noting $\hat{x}_3 = s_2 + \varsigma_d = s_2 + z_2 + \bar{\varsigma}_d$ yields:

$$\dot{s}_1 = \bar{A}_a(s_2 + z_2) - \frac{1}{m}\left[\bar{b}x_2 + \tilde{F}_f S_f(\dot{x}) + \tilde{d}_f\right] + \bar{A}_a \tilde{x}_3 - k_2 s_1 - \frac{h_1^2}{4\eta_1}s_1,\tag{29}$$

where $\tilde{b} = b - \hat{b}, \tilde{F}_f = F_f - \hat{F}_f, \tilde{d}_f = d_f - \hat{d}_f$.

Substituting Equation (22) into Equation (21) and noting $\hat{x}_4 = s_3 + \hat{\zeta}_d = s_3 + z_3 + \bar{\zeta}_d$ yields:

$$\dot{s}_2 = \frac{RT_s a}{V_a}(s_3 + z_3) - k_3 s_2. \tag{30}$$

Substituting Equation (25) into Equation (24) gives:

$$\dot{s}_3 = -k_4 s_3. \tag{31}$$

Therefore, the time derivative of $V_1 = V_{cs} + V_\theta$ is:

$$
\begin{aligned}
\dot{V}_1 = \quad & -k_1 e_1^2 - k_2 s_1^2 - k_3 s_2^2 - k_4 s_3^2 + e_1 s_1 + e_1 z_1 + \bar{A}_a(s_1 s_2 + s_1 z_2) + \frac{RT_s a}{V_a}(s_3 + z_3) \\
& + \gamma_1^{-1} \tilde{b}\left[\dot{\hat{b}} - \left(-\frac{\gamma_1}{m} x_2 s_1\right)\right] + \gamma_2^{-1} \tilde{F}_f\left[\dot{\hat{F}}_f - \left(-\frac{\gamma_2}{m} S_f(x_2)s_1\right)\right] + \gamma_3^{-1}\tilde{d}_f\left[\dot{\hat{d}}_f - \left(-\frac{\gamma_3}{m}s_1\right)\right] \\
& + s_1\left\{-\frac{h_1^2}{4\eta_1}s_1 - \frac{1}{m}\left[\bar{b}x_2 + \tilde{F}_f S_f(\dot{x}) + \tilde{d}_f\right] + \bar{A}_a \tilde{x}_3\right\}
\end{aligned} \tag{32}
$$

It has been proved in [36] that with the projection mapping, the following properties hold:

$$\tilde{b}\left[\text{Proj}_{\hat{b}}\left(-\frac{\gamma_1}{m}x_2 s_1\right) - \left(-\frac{\gamma_1}{m}x_2 s_1\right)\right] \le 0. \tag{33a}$$

$$\tilde{F}_f\left[\text{Proj}_{\hat{F}_f}\left(-\frac{\gamma_2}{m}S_f(x_2)s_1\right) - \left(-\frac{\gamma_2}{m}S_f(x_2)s_1\right)\right] \le 0. \tag{34b}$$

$$\tilde{d}_f\left[\text{Proj}_{\hat{d}_f}\left(-\frac{\gamma_3}{m}s_1\right) - \left(-\frac{\gamma_3}{m}s_1\right)\right] \le 0. \tag{33c}$$

According to Assumption 1–2, there exists a known function h_1 satisfies:

$$h_1(t) \ge \frac{1}{m}\left[|b_{max} - b_{min}||x_2| + |F_{fmax} - F_{fmin}||S_f(x_2)| + |d_{max} - d_{min}|\right] + \bar{A}_a|\rho_1|, \tag{34}$$

where ρ_1 is the bound of $\tilde{x}_3$ as proven in the above step. From the smoothed sliding mode control theory, the following inequality holds.

$$s_1\left\{-\frac{h_1^2}{4\eta_1}s_1 - \frac{1}{m}\left[\bar{b}x_2 + \tilde{F}_f S_f(\dot{x}) + \tilde{d}_f\right] + \bar{A}_a\tilde{x}_3\right\} \le \eta_1. \tag{35}$$

Substituting Equation (34) into Equation (32) yields

$$\dot{V}_1 \le -k_1 e_1^2 - k_2 s_1^2 - k_3 s_2^2 - k_4 s_3^2 + e_1 s_1 + e_1 z_1 + \bar{A}_a(s_1 s_2 + s_1 z_2) + \frac{RT_s a}{V_a}(s_3 + z_3) + \eta_1. \tag{36}$$

The following inequalities can be obtained by using Young's inequality.

$$
\begin{aligned}
e_1 s_1 &\le \frac{e_1^2}{2} + \frac{s_1^2}{2}, e_1 z_1 \le \frac{e_1^2}{2} + \frac{z_1^2}{2}, s_1 s_2 \le \frac{s_1^2}{2} + \frac{s_2^2}{2} \\
s_1 z_2 &\le \frac{s_1^2}{2} + \frac{z_2^2}{2}, s_2 s_3 \le \frac{s_2^2}{2} + \frac{s_3^2}{2}, s_2 z_3 \le \frac{s_2^2}{2} + \frac{z_3^2}{2}
\end{aligned} \tag{37}
$$

Substituting Equation (37) into Equation (36) gives:

$$
\begin{aligned}
\dot{V}_1 \le \quad & -(k_1 - 1)e_1^2 - (k_2 - \tfrac{1}{2} - \bar{A}_a)s_1^2 - (k_3 - \tfrac{\bar{A}_a}{2} - \tfrac{RT_s a}{V_a})s_2^2 - (k_4 - \tfrac{RT_s a}{2V_a})s_3^2 + \\
& \tfrac{1}{2}z_1^2 + \tfrac{\bar{A}_a}{2}z_2^2 + \tfrac{RT_s a}{2V_a}z_3^2 + \eta_1
\end{aligned} \tag{38}
$$

Following the similar analysis as done in [28], the following inequalities hold.

$$\dot{z}_1 + \frac{z_1}{\tau_1} \le \left|\dot{z}_1 + \frac{z_1}{\tau_1}\right| \le \xi_1, \quad \dot{z}_2 + \frac{z_2}{\tau_2} \le \left|\dot{z}_2 + \frac{z_2}{\tau_2}\right| \le \xi_2, \quad \dot{z}_3 + \frac{z_3}{\tau_3} \le \left|\dot{z}_3 + \frac{z_3}{\tau_3}\right| \le \xi_3, \tag{39}$$

where $\xi_1(e_1, s_1, z_1, x_d, \dot{x}_d, \ddot{x}_d)$, $\xi_2(e_1, s_1, s_2, z_1, z_2, x_d, \dot{x}_d, \ddot{x}_d)$, and $\xi_2(e_1, s_1, s_2, s_3, z_1, z_2, z_3, x_d, \dot{x}_d, \ddot{x}_d)$ are three continuous functions. Due to the fact that the set Ω_0 is compact in R^3 and the set $\Omega_1 = \left\{ [e_1, s_1, s_2, s_3, z_1, z_2, z_3]^{\mathrm{T}} : e_1^2 + s_1^2 + s_2^2 + s_3^2 + z_1^2 + z_2^2 + z_3^2 \leq 2B_1 \right\}$ (where $B_1 > 0$) is compact in R^7, $\Omega_0 \times \Omega_1$ is thus compact in R^{10}. Therefore, ξ_1, ξ_2, and ξ_3 have maximums on $\Omega_0 \times \Omega_1$. Let M_1, M_2, and M_3 be the maximums of ξ_1, ξ_2, and ξ_3 on $\Omega_0 \times \Omega_1$, respectively, one can obtain that:

$$z_1 \dot{z}_1 + \frac{z_1^2}{\tau_1} \leq M_1 |z_1|, \quad z_2 \dot{z}_2 + \frac{z_2^2}{\tau_2} \leq M_2 |z_2|, \quad z_3 \dot{z}_3 + \frac{z_3^2}{\tau_3} \leq M_3 |z_3|. \tag{40}$$

With the use of Young's inequality, one can obtain that:

$$z_1 \dot{z}_1 \leq -\frac{z_1^2}{\tau_1} + \frac{z_1^2}{2} + \frac{M_1^2}{2}, \quad z_2 \dot{z}_2 \leq -\frac{z_2^2}{\tau_2} + \frac{z_2^2}{2} + \frac{M_2^2}{2}, \quad z_3 \dot{z}_3 \leq -\frac{z_3^2}{\tau_3} + \frac{z_3^2}{2} + \frac{M_3^2}{2}. \tag{41}$$

Thus, the following inequalities can be obtained.

$$\dot{V}_{cz} \leq -\frac{1}{\tau_1} z_1^2 - \frac{1}{\tau_2} z_2^2 - \frac{1}{\tau_3} z_3^2 + \frac{1}{2} z_1^2 + \frac{1}{2} M_1^2 + \frac{1}{2} z_2^2 + \frac{1}{2} M_2^2 + \frac{1}{2} z_3^2 + \frac{1}{2} M_3^2. \tag{42}$$

Thus, the following inequalities can be obtained.

$$\begin{aligned}
\dot{V}_1 \leq \quad & -(k_1 - 1)e_1^2 - (k_2 - \tfrac{1}{2} - \overline{A}_a)s_1^2 - (k_3 - \tfrac{\overline{A}_a}{2} - \tfrac{RT_s a}{V_a})s_2^2 - (k_4 - \tfrac{RT_s a}{2V_a})s_3^2 - \\
& (\tfrac{1}{\tau_1} - 1)z_1^2 - (\tfrac{1}{\tau_2} - \tfrac{1}{2} - \tfrac{\overline{A}_a}{2})z_2^2 - (\tfrac{1}{\tau_3} - \tfrac{1}{2} - \tfrac{RT_s a}{2V_a})z_3^2 + \tfrac{1}{2}M_1^2 + \tfrac{1}{2}M_2^2 + \tfrac{1}{2}M_3^2 + \eta_1
\end{aligned} \tag{43}$$

By choosing $k_1 > 1$, $k_2 > \frac{1}{2} + \overline{A}_a$, $k_3 > \frac{\overline{A}_a}{2} + \frac{RT_s a}{V_a}$, $k_4 > \frac{RT_s a}{2V_a}$, $\tau_1 < 1$, $\tau_2 < \frac{2}{\overline{A}_a + 1}$, and $\tau_3 < \frac{2V_a}{RT_s a + V_a}$, one has:

$$\gamma = \min\left\{ k_1 - 1, k_2 - \frac{1}{2} - \overline{A}_a, k_3 - \frac{\overline{A}_a}{2} - \frac{RT_s a}{V_a}, k_4 - \frac{RT_s a}{2V_a}, \frac{1}{\tau_1} - 1, \frac{1}{\tau_2} - \frac{1}{2} - \frac{\overline{A}_a}{2}, \frac{1}{\tau_3} - \frac{1}{2} - \frac{RT_s a}{2V_a} \right\} > 0 \tag{44}$$

Then, it can be readily obtained that the following inequality holds.

$$\dot{V} \leq -2\gamma V + \eta, \tag{45}$$

where $\eta = \frac{1}{2}M_1^2 + \frac{1}{2}M_2^2 + \frac{1}{2}M_3^2 + \eta_1$. $\square$

Clearly, if $\gamma > \eta/2B_1$, then $\dot{V} \leq 0$ on set Ω_1, thus V will remain $V(t) \leq B_1$ for all t, provided that the initial conditions satisfy $V(0) \leq B_1$. Therefore, the errors $e_1, s_1, s_2, s_3, z_1, z_2, z_3$ are bounded. Hence, one can conclude that the closed loop system is uniformly and ultimately bounded.

4. Experimental Results

In this section, the proposed controller was implemented for the servo control of the pneumatic servo system as shown in Figure 2. The cylinder forward chamber was controlled by the valve through a 10 m long transmission line with 4 mm inside diameter, while the pressure in the return chamber was maintained at about 3 bar by utilizing a tank. A gel phantom was used to perform the following needle insertion experiments. The system physical parameters and the parameters of the controller are given in Table 1.

Table 1. Parameters of the system and the proposed controller.

Symbol	Value	Unit
m	0.32	kg
A_a	7.854×10^{-5}	m^2
A_b	6.597×10^{-4}	m^2
A_r	1.257×10^{-5}	m^2
R	287	N m/(kg K)
T_s	300	K
p_s	6×10^5	Pa
V_{a0}	6×10^{-7}	m^3
L_c	0.125	m
V_l	1.257×10^{-4}	m^3
C_d	$1.099 - 0.1075 \times \frac{p_u}{p_d}$	
p_r	0.29	
A_l	1.257×10^{-5}	m^2
D_l	0.004	m
L_l	10	m
μ	1.79×10^{-5}	N s/m^2
p_0	1×10^5	Pa
$\hat{b}(0), b_{min}, b_{max}$	9, 0, 15	N s/m
$\hat{F}_f(0), \hat{F}_{fmin}, \hat{F}_{fmax}$	6, 0, 20	N
$\hat{d}_f(0), d_{min}, d_{max}$	0, −10, 10	N
k_1, τ_1	40, 0.5	
k_2, h_1, η_1, τ_2	50, 40, 0.4, 0.75	
k_3, τ_3	50, 1.2	
k_4	90	
$\gamma_1, \gamma_2, \gamma_3$	100, 10, 10	

The same control algorithm as the proposed controller, but in which the long transmission line was characterized as a part of the cylinder's dead volume and the chamber pressure was directly measured, was first tested for tracking a sinusoidal trajectory and a smooth square trajectory. As shown in Figure 4, the performance was poor for practical application, which indicates the need for an advanced method to address the issue of long transmission line. The performance of the proposed controller was tested on three types of reference trajectories: A 0.5 Hz sinusoidal signal, a smooth square signal, and a 4 s periodic signal. Figure 5 shows the response of the system, and the tracking errors are presented in Figure 6. The maximum absolute tracking errors ($\max_t\{|e_1|\}$) were about 2.51 mm, 2.68 mm, and 1.81 mm, while the final steady-state tracking errors ($\max_{t\geq 5}\{|e_1|\}$) were about 0.96 mm, 1.05 mm, and 1.39 mm. As seen, the proposed controller could significantly improve the system performance, which indicates the effectiveness of the proposed long transmission line compensation method. However, it should be noted that these needle insertion experiments were performed in a gel phantom, whose resisting force behavior was simpler than real soft tissue. The interaction between needle and soft tissue is very complex and may decrease the needle insertion precision in practical. Since the current research focused on addressing the issue of long transmission line and realizing high accuracy control of a pneumatic actuator with a pressure observer, modeling of needle-tissue interaction and further improving needle insertion precision would be the subject of the next phase of this research.

The process of parameter and disturbance estimations is shown in Figure 7. It can be seen that the estimates converged quickly, and the tracking error improvement could be achieved within several seconds. Comparisons between the observed chamber pressure and the measured one are shown in Figure 8. As can be seen, the estimated chamber pressure was close to its actual value, which indicates the effectiveness of the proposed pressure observer. However, the estimation error during the charging process was slightly bigger than the one during the discharging process. This might be due to the fact that charging process was close to adiabatic and the discharging process was close to isothermal.

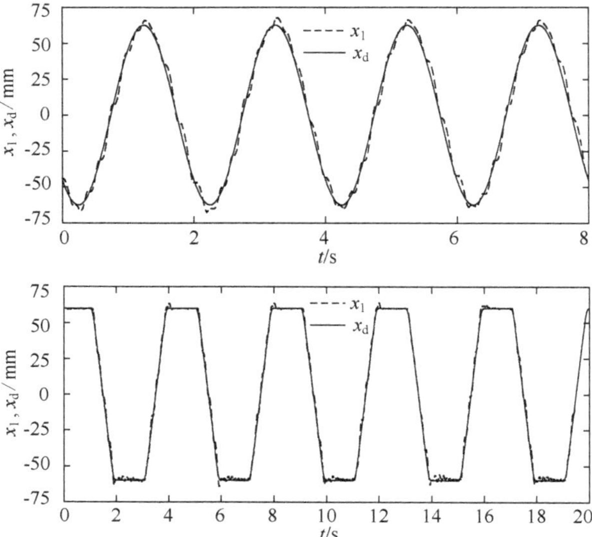

Figure 4. Tracking response of the same control algorithm as the proposed controller but without using long transmission line compensation.

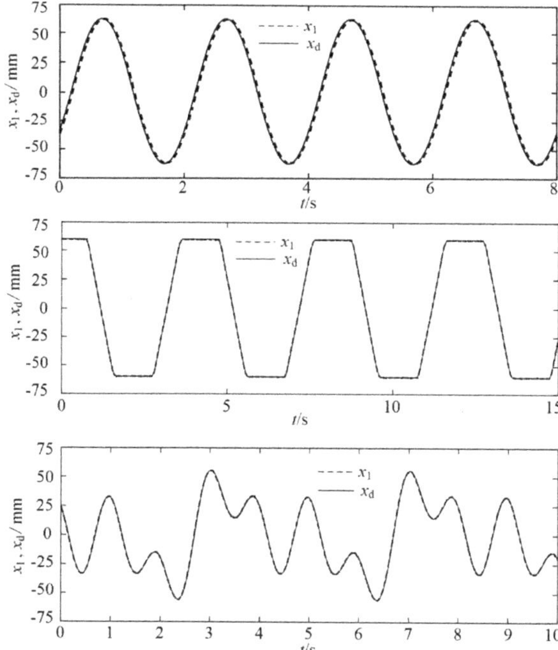

Figure 5. Tracking response of the proposed controller for three different types of trajectories.

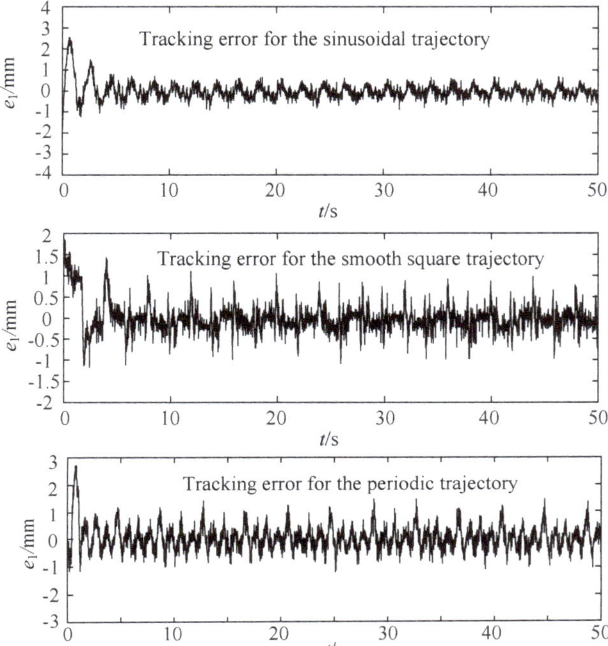

Figure 6. Tracking errors of the proposed controller for three different types of trajectories.

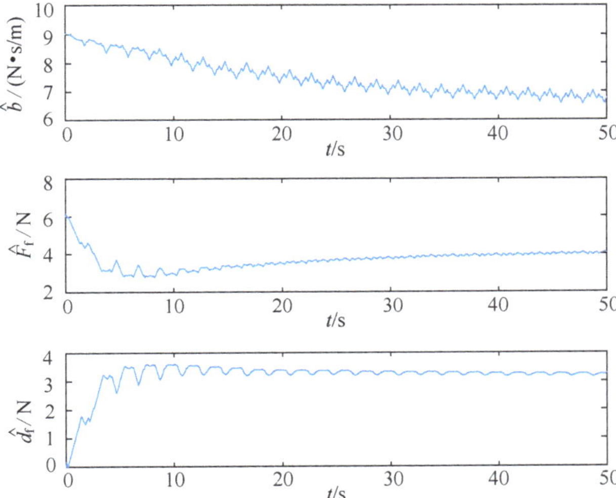

Figure 7. Parameter and disturbance estimations of the proposed controller when the reference is a 0.5 Hz sinusoidal trajectory.

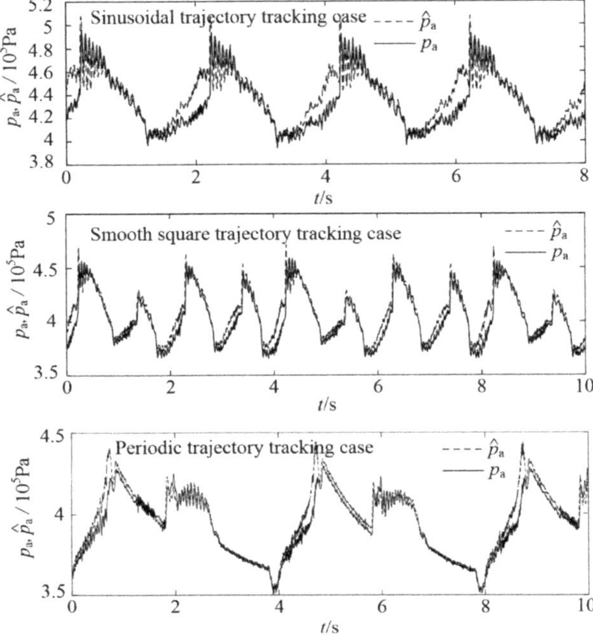

Figure 8. Actual and observed chamber pressure for the system tracking three different types of references.

Experiments were also conducted to verify the performance robustness of the proposed controller. To simulate a sudden disturbance acting on the system, a big step signal was added to the output of the position sensor at $t = 22$ s, and removed 10 s later. Figure 9 shows the control accuracy of the proposed controller in this scenario for tracking a sinusoidal trajectory with a frequency of 0.5 Hz and amplitude of 62.5 mm. It can be seen that the system performance did not deteriorate except the transient spikes when the sudden disturbance was added or removed.

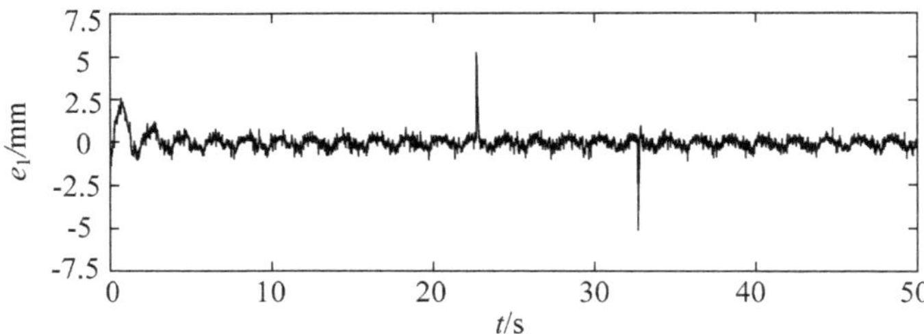

Figure 9. Tracking error of the proposed controller for 0.5 Hz sinusoidal trajectory with disturbance.

5. Conclusions

In this paper, the precise motion control of a MRI compatible 1-DOF pneumatic servo system was considered. The long transmission line was characterized as an intermediate chamber connected between the valve and the cylinder in series, and a nonlinear first order system was used to approximate its dynamics. Simultaneously, a globally stable pressure observer was developed to estimate the chamber pressure. Based on the model of the long transmission line and the pressure observer, a pressure observer based adaptive dynamic surface controller was developed and the stability of the closed-loop system was proved via the Lyapunov method. In contrast to most of the existing nonlinear controllers synthesized with the backstepping method, by employing the dynamic surface control technique, the proposed controller could cope with the problem of "explosion of complexity", since the repeated differentiation of virtual controls was no longer required. The experimental results confirmed that the proposed controller was effective and had good performance robustness to sudden disturbances, thus enabling future application in pneumatically actuated MRI-compatible robots. However, precise motion control of pneumatic actuator did not necessarily lead to precise position control needle tip. Modeling of the interaction between needle and soft tissue, and incorporating a more accurate needle insertion force model in the controller design is an essential requirement for practical robot-assisted needle insertion. Thus, these issues will be explored further in the next phase of this research.

Author Contributions: D.M. and B.L. proposed the pressure observe based dynamic surface controller for precise position tracking control of the MRI-compatible pneumatic servo system. A.L. designed the experiments. J.Y. and Q.L. performed the experiments and analyzed the data. D.M. and A.L. wrote the paper.

Funding: This research was funded by [the Fundamental Research Funds for the Central Universities] grant number [2015XKMS020], [the National Natural Science Foundation of China] grant number [51505474], [the China Postdoctoral Science Foundation] grant number [2016T90520], and [a Project Funded by the Priority Academic Program Development of Jiangsu Higher Education Institutions]. And the APC was funded by the Fundamental Research Funds for the Central Universities.

Acknowledgments: This work was supported by the National Natural Science Foundation of China (Grant No. 51505474), the Fundamental Research Funds for the Central Universities (Grant No. 2015XKMS020), the China Postdoctoral Science Foundation (Grant No. 2016T90520), and a Project Funded by the Priority Academic Program Development of Jiangsu Higher Education Institutions.

Conflicts of Interest: The authors declare no conflict of interest.

References

1. Fischer, G.; Iordachita, I.; Csoma, C.; Tokuda, J.; DiMaio, S.; Tempany, C.; Hata, N.; Fichtinger, G. MRI-compatible pneumatic robot for transperineal prostate needle placement. *IEEE/ASME Trans. Mechatron.* **2008**, *13*, 295–305. [CrossRef] [PubMed]

2. Stoianovici, D.; Jun, C.; Lim, S.; Li, P.; Petrisor, D.; Fricke, S.; Sharma, K.; Cleary, K. Multi-image compatible, MR safe, remote center of motion needle-guide robot. *IEEE Trans. Biomed. Eng.* **2018**, *65*, 165–177. [CrossRef] [PubMed]

3. Stoianovici, D.; Kim, C.; Srimathveeravalli, G.; Sebrecht, P.; Petrisor, D.; Coleman, J.; Solomon, S.; Hricak, H. MRI-Safe Robot for Endorectal Prostate Biopsy. *Trans. Mechatron.* **2014**, *19*, 1289–1299. [CrossRef]

4. Stoianovici, D.; Patriciu, A.; Petrisor, D.; Mazilu, D.; Kavoussi, L. A new type of motor: Pneumatic step motor. *IEEE/ASME Trans. Mechatron.* **2007**, *12*, 98–106. [CrossRef] [PubMed]

5. Zemiti, N.; Bricault, I.; Fouard, C.; Sanchez, B.; Cinquin, P. LPR: A CT and MR-compatible puncture robot to enhance accuracy and safety of image-guided interventions. *IEEE/ASME Trans. Mechatron.* **2008**, *13*, 306–315. [CrossRef]

6. Schouten, M.; Bomers, J.; Yakar, D.; Huisman, H.; Rothgang, E.; Bosboom, D.; Scheenen, T.; Misra, S.; Futterer, J. Evaluation of a robotic technique for transrectal MRI-guided prostate biopsies. *Eur. Radiol.* **2012**, *22*, 476–483. [CrossRef] [PubMed]

7. Franco, E.; Brujic, D.; Rea, M.; Gedroyc, W.; Ristic, M. Needle-guiding robot for laser ablation of liver tumors under MRI guidance. *IEEE/ASME Trans. Mechatron.* **2016**, *21*, 931–944. [CrossRef]

8. Yang, B.; Tan, U.; McMillan, A.; Gullapalli, R.; Desai, J. Design and control of a 1-DOF MRI-compatible pneumatically actuated robot with long transmission lines. *IEEE/ASME Trans. Mechatron.* **2011**, *16*, 1040–1048. [CrossRef] [PubMed]

9. Yang, B.; Roys, S.; Tan, U.; Philip, M.; Richard, H.; Gullapalli, R.; Desai, J. Design, development, and evaluation of a master–slave surgical system for breast biopsy under continuous MRI. *Int. J. Robot. Res.* **2014**, *33*, 616–630. [CrossRef]

10. Van den Bosch, M.; Moman, M.; van Vulpen, M.; Battermann, J.; Duiveman, E.; van Schelven, L.; de Leeuw, H.; Lagendijk, J.; Moerland, M. MRI-guided robotic system for transperineal prostate interventions: Proof of principle. *Phys. Med. Biol.* **2010**, *55*, 133–140. [CrossRef]

11. Chen, Y.; Squires, A.; Seifabadi, R.; Xu, S.; Agarwal, H.; Bernardo, M.; Pinto, P.; Choyke, P.; Wood, B.; Tse, Z. Robotic system for MRI-guided focal laser ablation in the prostate. *IEEE/ASME Trans. Mechatron.* **2017**, *22*, 107–114. [CrossRef] [PubMed]

12. Chen, Y.; Xu, S.; Squires, A.; Seifabadi, R.; Turkbey, I.; Pinto, P.; Choyke, P.; Wood, B.; Tse, Z. MRI-guided robotically assisted focal laser ablation of the prostate using canine cadavers. *IEEE/ASME Trans. Mechatron.* **2018**, *65*, 1434–1442. [CrossRef] [PubMed]

13. Melzer, A.; Gutmann, B.; Remmele, T.; Wolf, R.; Lukoscheck, A.; Bock, M.; Bardenheuer, H.; Fischer, H. INNOMOTION for percutaneous image-guided interventions. *IEEE Eng. Med. Biol. Mag.* **2008**, *27*, 66–73. [CrossRef] [PubMed]

14. Jiang, S.; Feng, W.; Lou, J.; Yang, Z.; Liu, J.; Yang, J. Modeling and control of a five-degrees-of-freedom pneumatically actuated magnetic resonance-compatible robot. *Int. J. Med. Robot. Comput. Assist. Surg.* **2014**, *10*, 170–179. [CrossRef] [PubMed]

15. Comber, D.; Barth, E.; Webster, R. Design and control of a magnetic resonance compatible precision pneumatic active cannula robot. *J. Med. Dev.* **2014**, *8*, 1011003-1–1011003-7. [CrossRef]

16. Richer, E.; Hurmuzlu, Y. A high performance pneumatic force actuator system: Part 1—nonlinear mathematical model. *J. Dyn. Syst. Meas. Control* **2000**, *122*, 416–425. [CrossRef]

17. Richer, E.; Hurmuzlu, Y. A high performance pneumatic force actuator system: Part II—nonlinear controller design. *J. Dyn. Syst. Meas. Control* **2000**, *122*, 426–434. [CrossRef]

18. Turkseven, M.; Ueda, J. An asymptotically stable pressure observer based on load and displacement sensing for pneumatic actuators with long transmission lines. *IEEE/ASME Trans. Mechatron.* **2017**, *22*, 681–692. [CrossRef]

19. Li, J.; Kawashima, K.; Fujita, T.; Kagawa, T. Control design of a pneumatic cylinder with distributed model of pipelines. *Precis. Eng.* **2013**, *37*, 880–887. [CrossRef]

20. Krichel, S.; Sawodny, O. Non-linear friction modeling and simulation of long pneumatic transmission lines. *Math. Comput. Model. Dyn. Syst.* **2014**, *20*, 23–44. [CrossRef]

21. Turkseven, M.; Ueda, J. Model-based force control of pneumatic actuators with long transmission lines. *IEEE/ASME Trans. Mechatron.* **2018**, *23*, 1292–1302. [CrossRef]

22. Pandian, S.; Takemura, F.; Hayakawa, Y.; Kawamura, S. Pressure observer-controller design for pneumatic cylinder actuators. *IEEE/ASME Trans. Mechatron.* **2002**, *7*, 490–499. [CrossRef]

23. Wu, J.; Goldfarb, M.; Barth, E. On the observability of pressure in a pneumatic servo actuator. *J. Dyn. Syst. Meas. Control* **2004**, *126*, 921–924. [CrossRef]

24. Bigras, P. Sliding-mode observer as a time-variant estimator for control of pneumatic systems. *J. Dyn. Syst. Meas. Control* **2005**, *127*, 499–502. [CrossRef]

25. Gulati, N.; Barth, E. A globally stable, load-independent pressure observer for the servo control of pneumatic actuators. *IEEE/ASME Trans. Mechatron.* **2009**, *14*, 295–306. [CrossRef]

26. Langjord, H.; Kassa, G.; Johansen, T. Adaptive nonlinear observer for electropneumatic clutch actuator with position sensor. *IEEE Trans. Control Syst. Technol.* **2012**, *20*, 1033–1039. [CrossRef]

27. Driver, T.; Shen, X. Pressure estimation-based robust force control of pneumatic actuators. *Int. J. Fluid Power* **2013**, *14*, 37–45. [CrossRef]

28. Swaroop, D.; Hedrick, J.K.; Yip, P.P.; Gerdes, J.C. Dynamic surface control for a class of nonlinear systems. *IEEE Trans. Autom. Control* **2000**, *45*, 1893–1899. [CrossRef]

29. Kheowree, T.; Kuntanapreeda, S. Adaptive dynamic surface control of an electrohydraulic actuator with friction compensation. *Asian J. Control* **2015**, *17*, 855–867. [CrossRef]

30. Peng, Z.; Wang, D.; Chen, Z.; Hu, X.; Lan, W. Adaptive dynamic surface control for formations of autonomous surface vehicles with uncertain dynamics. *IEEE Trans. Control Syst. Technol.* **2013**, *21*, 513–520. [CrossRef]

31. Huang, J.; Ri, S.; Liu, L.; Wang, Y.; Kim, Y.; Pak, G. Nonlinear disturbance observer-based dynamic surface control of mobile wheeled inverted pendulum. *IEEE Trans. Control Syst. Technol.* **2015**, *23*, 2400–2407. [CrossRef]

32. Wu, J.; Huang, J.; Wang, Y.; Xing, K. Nonlinear disturbance observer-based dynamic surface control for trajectory tracking of pneumatic muscle system. *IEEE Trans. Control Syst. Technol.* **2014**, *22*, 440–455. [CrossRef]

33. Wang, M.; Ren, X.; Chen, Q.; Wang, S.; Gao, X. Modified dynamic surface approach with bias torque for multi-motor servomechanism. *Control Eng. Pract.* **2016**, *50*, 57–68. [CrossRef]

34. Li, A.; Meng, D.; Lu, B.; Li, Q. Nonlinear cascade control of single-rod pneumatic actuator based on an extended disturbance observer. *J. Cent. South Univ.* **2019**, *26*, 1637–1648. [CrossRef]
35. Meng, D.; Tao, G.; Chen, J.; Ban, W. Modeling of a pneumatic system for high-accuracy position control. In Proceedings of the International Conference on Fluid Power and Mechatronics, Beijing, China, 17–20 August 2011; pp. 505–510.
36. Goodwin, G.; Mayne, D. A parameter estimation perspective of continuous time model reference adaptive control. *Automatica* **1989**, *23*, 57–70. [CrossRef]

Article

Tip Estimation Method in Phantoms for Curved Needle Using 2D Transverse Ultrasound Images

Zihao Li [1], Shuang Song [1,*], Li Liu [2,*] and Max Q.-H. Meng [2]

[1] School of Mechanical Engineering and Automation, Harbin Institute of Technology, Shenzhen 518055, China; lzh_lejou@163.com

[2] Robotics, Perception and Artificial Intelligence Lab, The Chinese University of Hong Kong, N.T., Hong Kong, China; qhmeng@ee.cuhk.edu.hk

* Correspondence: songshuang@hit.edu.cn (S.S.); liliu@cuhk.edu.hk (L.L.)

Received: 8 October 2019; Accepted: 27 November 2019; Published: 5 December 2019

Abstract: Flexible needles have been widely used in minimally invasive surgeries, especially in percutaneous interventions. Among the interventions, tip position of the curved needle is very important, since it directly affects the success of the surgeries. In this paper, we present a method to estimate the tip position of a long-curved needle by using 2D transverse ultrasound images from a robotic ultrasound system. Ultrasound is first used to detect the cross section of long-flexible needle. A new imaging approach is proposed based on the selection of numbers of pixels with a higher gray level, which can directly remove the lower gray level to highlight the needle. After that, the needle shape tracking method is proposed by combining the image processing with the Kalman filter by using 3D needle positions, which develop a robust needle tracking procedure from 1 mm to 8 mm scan intervals. Shape reconstruction is then achieved using the curve fitting method. Finally, the needle tip position is estimated based on the curve fitting result. Experimental results showed that the estimation error of tip position is less than 1 mm within 4 mm scan intervals. The advantage of the proposed method is that the shape and tip position can be estimated through scanning the needle's cross sections at intervals along the direction of needle insertion without detecting the tip.

Keywords: needle tracking; tip estimation; transverse ultrasound images; percutaneous interventions

---

1. Introduction

With the help of the beveled-tip needle, percutaneous interventions and therapies have been widely involved in current clinical procedures, such as brachytherapy [1,2], tissue biopsy [3,4], and drug delivery [5,6]. In intervention procedures, less needle misplacement will lead to a more reliable treatment and a more accurate medical practice. According to the clinical studies [7,8], the needle is easy to be deflected, which will cause needle tip misplacement and may lead to unsafe procedures. Due to the needle-tissue interaction, improper insertion force or physiological motions, such as breathing may cause targets or obstacles to be unstable, which will lead to an unexpected error. To address the challenge, real-time feedback is highly required. Usually, medical imaging devices are used, such as ultrasound (US) [9], computerized tomography (CT) [10,11], or magnetic resonance imaging (MRI) [2,12]. Generally, the image-guided percutaneous interventions are conducted with the use of CT or MRI. However, ultrasound-guided procedures are more attractive due to their advantages such as none ionizing radiations and real-time detection.

Many studies for the guidance of the needle during the insertion operation have been conducted with the help of ultrasound devices, and 2D ultrasound images are quite general to use, especially for the sagittal one (shown in Figure 1). Elif et al. proposed to use circular Hough transform to locate the needle tip accurately, even when the imaging is out-of-plane [13]. Kaya et al. localized the needle axis and estimated the needle tip by using a Gabor Filter in sagittal US images [14]. To execute in real-time, they improved the processing time by applying the bin packing method [15]. Recently, a template-based tracking method with the efficient second-order minimization optimization method has been used to track the needle [16]. In recent studies, more and more novel ideas have been used to locate the needle and evaluate its tip to sagittal US images, such as the use of signal attenuation maps [17], convolution neural networks (CNN) [18], and maximum likelihood estimation sample consensus (MLESAC) method [19]. However, a demerit of using sagittal US images is that out-of-plane bending of the needle cannot be detected. Therefore, the methods applied on sagittal US images are not suitable for the needle which may be deflected by the inevitable factors, especially for long needles.

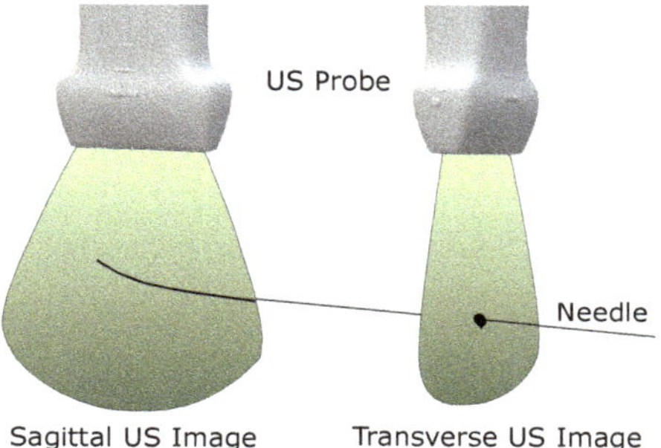

Figure 1. Two methods by using 2D ultrasound for detection.

An alternate solution for this problem is to use a 3D US image, which has been widely studied in recent researches. Yue et al. used a RANSAC method to detect the needle in a 3D US situation and Kalman filter has been used to reduce the error [20]. Chatelain et al. used the particle filtering to track a robot-guided flexible needle by using 3D US [21]. In addition, a convolutional neural network with conventional image processing techniques has also been used to track and detect the needle [22] and a naive Bayesian classifier was used to localize the needle among 3D US volume voxels [23]. However, the large 3D US volumetric dataset would make it difficult to obtain and process the online data.

Due to the above disadvantages, sagittal US images and 3D US volume are not suitable for a long flexible needle. To locate the needle accurately, methods that use transverse US images (shown in Figure 1) have been used successfully in some studies. For example, Vrooijink et al. [24] present a method to track the flexible needle during the insertion into a gelatin tissue by using 2D US images perpendicular to its needle tip. However, the background is pure without noise, which makes it impractical. Waine et al. [25–27] focus on the research about the needle insertion, in permanent prostate brachytherapy (PPB), where needles are typically 200 mm and easily to be deflected, indicating the fact that the rectum limits the movement of US probe. As a result, it is hard to acquire the sagittal images of the curved needle to observe the deflection during needle insertion. This is because the sagittal method has a strong relationship to the movement of the US probe when the needle is out of the view-field of the US images, and this movement maybe deforms the prostate as well as affects the needle target. For a deflected needle, the transverse US image is a better choice for its detection. Compared to the sagittal images, the transverse US images are easy to be acquired when the US probe scanning along the needle, no matter how much the needle is curved.

In this paper, we present a method to track a long-curved needle from the 2D transverse US images and estimate its tip for the guidance of needle insertion. Ultrasound is first used to detect the cross-sections of the long-flexible needle (shown in Figure 2 STEP 1). The needle shape tracking method combined needle detection with Kalman filter develops an accurate location and a robust tracking procedure with scan intervals from 1 mm to 8 mm (shown in Figure 2 STEP 2). Unlike the previous study [26], the 3D needle positions obtained from 2D US images and optical tracking systems have been used in KF for the precise location. The curve fitting method is then used to achieve the shape reconstruction and the needle tip position is estimated based on its length and the curve fitting result (shown in Figure 2 STEP 3). The advantage of the proposed method is that the shape and tip position can be estimated through scanning the needle's cross-sections at intervals along the direction of needle insertion without detecting the tip. Besides, a novel histogram method is introduced to detect the needle in image processing, which can improve the needle localization under the effect of needle comet tail and the poor reflection, despite of the abrupt intensity changes. In addition, a robotic ultrasound system (RUS) [28] is built to evaluate the proposed needle tip estimation method. Results showed that the estimation of the tip position is less than 1 mm with 4 mm scan intervals.

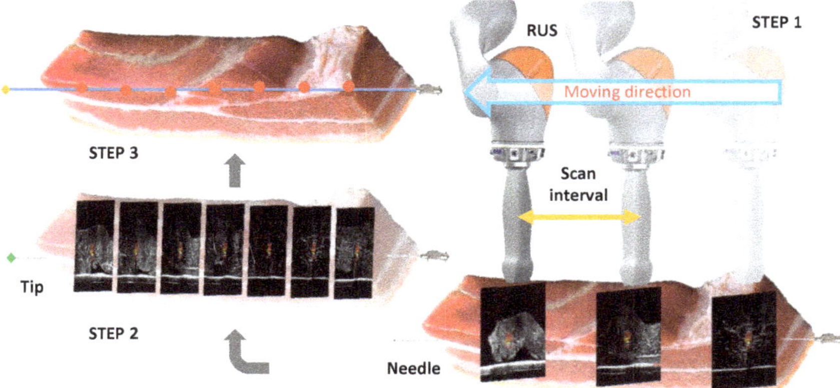

Figure 2. The proposed tip estimation method. STEP 1: 2D transverse US images with needle cross-sections are collected by using RUS; STEP 2: Needle cross-sections are detected and tracked in the successive US images; STEP 3: Needle shape is constructed and its tip is estimated.

The rest of this paper is organized as follows. The proposed methods will be introduced in detail in Section 2. Section 3 intends to represent the experimental setup and the results. Finally, the discussion and conclusions are detailed in Section 4.

2. Materials and Methods

The proposed needle tip estimation method in successive transverse US images can be divided into three stages: needle detection, needle shape tracking, and tip estimation. The processing diagram are shown in the Figure 3. The needle location will be manually selected as an initial region of interest (ROI) by the binary method in the first US image. After that, the prediction of needle position from a Kalman filter can be transformed in the transverse US images. At the same time, the next needle position in this ROI can be found through the histogram method. The KF is then updated for the current precise needle position and prediction of the next position. After all the cross-sections of the needle have been collected, the needle shape can be fitted by the curve fitting method (part C in Figure 3). Finally, the position of the needle tip can be estimated from the curve fitting result based on the length of the needle. In this study, the ROI is set as a square window with a length of three times larger than the needle diameter and its center represents the needle position. Needle detection (part B in Figure 3) is mainly about image processing, while the Kalman filter is used for needle shape tracking (part A in Figure 3).

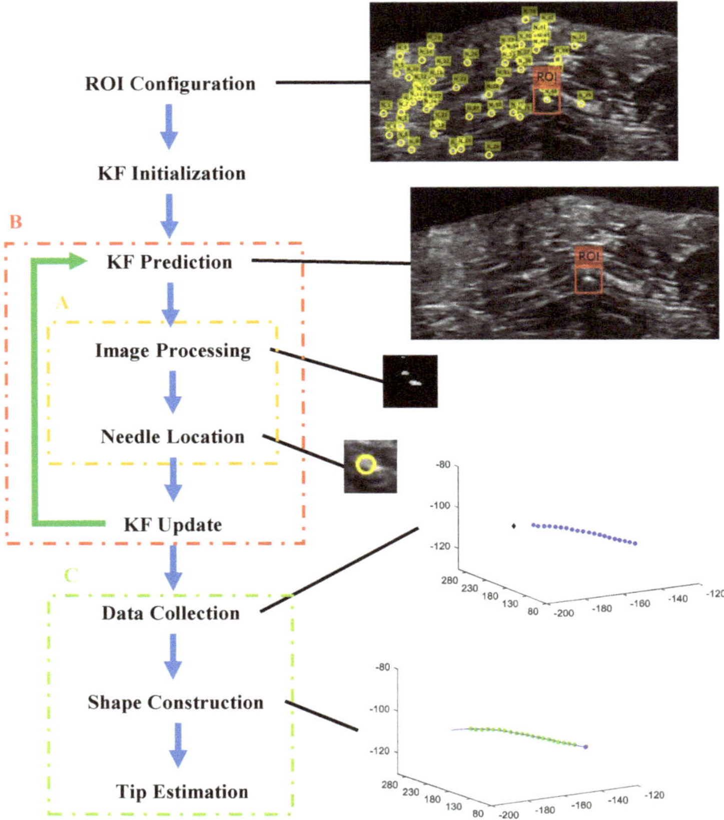

Figure 3. The pipeline of curved needle tip estimation. As the ROI configuration finished, needle tracking with needle detection begins step by step. Part A shows the needle tracking by KF, while Part B shows the needle detection to locate the needle tip. Part C shows the shape construction and tip estimation.

2.1. Needle Detection

Needle detection is mainly for identifying the cross-section of the needle in the US images by using the binary method and the histogram method. As the ultrasound is quite sensitive to the metal, the needle-inside area can be brighter than others. A binary method [26] is first used to select ROI in the first image and then a histogram method is used to locate the needle despite the US shadows and poor reflection.

Binary method intends to strengthen the contrast of brightness to highlight the brighter area to select them. This method is used for the initialization which supposes to find the candidates in the first image. It includes intensity normalization, background reinforcement, and brightness enhancement. The center of the area is the location of the needle. During the experiment, a histogram method is proposed to find the needle accurately. The histogram method contains intensity normalization and background reinforcement. The histogram method tends to find an area of high-intensity pixels, which intends to find the upper face of the needle and then locate the needle based on the diameter. The background reinforcement part can be described as follows:

$$\min \quad I_t$$
$$\text{s.t.} \quad \sum_{I_t}^{255} n_{I_j} \leq M\delta \tag{1}$$
$$I_t \in [0, 255]$$

where I_j is the gray level of the pixel and n_{I_j} is the number of pixels which have the gray level of I_j. M is the size of ROI, and δ is the manually selected parameter to limit the bright pixels. In this work, M is a square of 45×45, and δ is set to 0.08 based on empirical tests.

There are two unexpected situations that may affect the position accuracy, namely the comet tail and the poor reflection. The comet tail will affect the size of the needle area and usually lead to a larger area than the actual size (Figure 4a). On the contrary, the poor reflection makes the needle area look much smaller in the image (Figure 4b). Therefore, the accurate location should be intended to eliminate the effect of shadows and poor reflection. In Figure 4, there are two examples which are used the two methods relatively. As the example shown in Figure 4a, the noise (yellow circle in the histogram of ROI) may probably be concerned as the needle while the needle is just concerned about a few pixels (red circle in histogram of ROI). The two methods can both filter the noise and locate the needle correctly. However, in ROI configuration (shown in Figure 3), the histogram method intends to find more candidates than the binary method, since the former focuses on the higher intensity pixels while the latter focuses on the area and intensity. Therefore, the binary method is more feasible in ROI configuration.

However, in the case of poor reflection, the needle displays a little in the image, and the area of the needle is smaller than the expected. Because the needle would reflect as long as the image gain is high enough or the sound power is big enough, it would reveal apparently compared to its surroundings. An example of the histogram of ROI is shown in Figure 4b, the red circle represents the upper surface of the needle. Moreover, the ROI square is darker than the one in Figure 4a, while the settings of the ultrasound are the same in Figure 4. From the figure, the histogram method seems to be more accurate than the binary method. In fact, the error of two methods, in this case, is 0.34 mm with 1.2 mm diameter of the needle. It is not that obvious to judge the accuracy. In this study, we use the histogram method during the experiment.

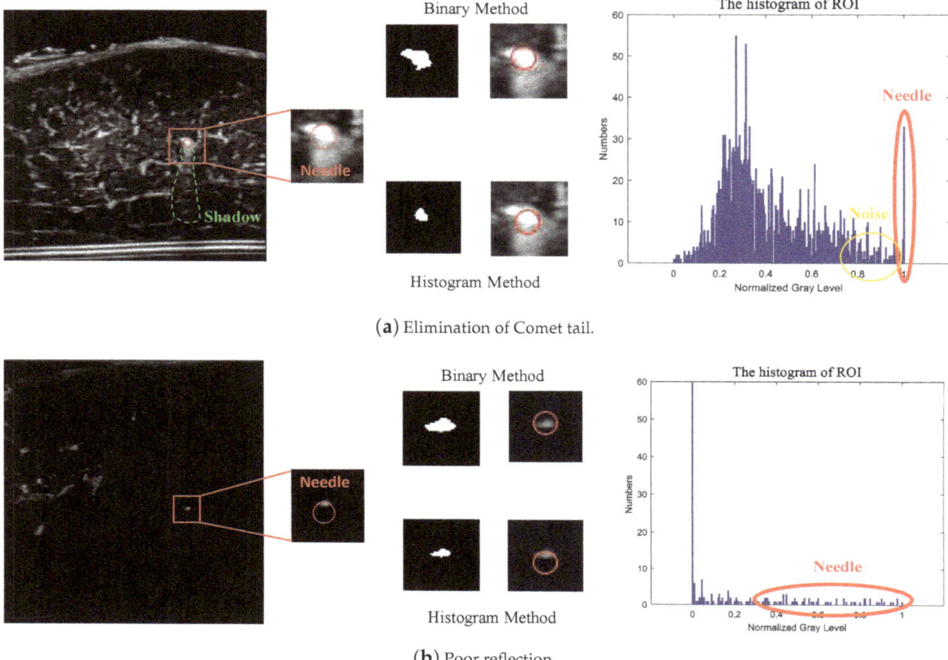

(**a**) Elimination of Comet tail.

(**b**) Poor reflection.

Figure 4. Two cases may generate the errors: (**a**) the comet tail of needle with binary method process, histogram method process and the histogram of ROI; (**b**) the poor reflection of the needle with histogram method process and the histogram of ROI.

2.2. Needle Shape Tracking

As indicated in previous researches [20,29,30], Kalman filter has been successfully used for tracking needle in the successive ultrasound images. In this study, the Kalman filter is used to improve the estimation of the needle location in successive frames. As shown in Figure 5, the applied Kalman filter has two processes, prediction and update. The prediction stage intends to locate the needle position previously and set the ROI (red and yellow square in Figure 5) to find the needle precisely with a small window, which is supposed to reduce the computation. The update stage is the result of the needle position after the measurement from the histogram method.

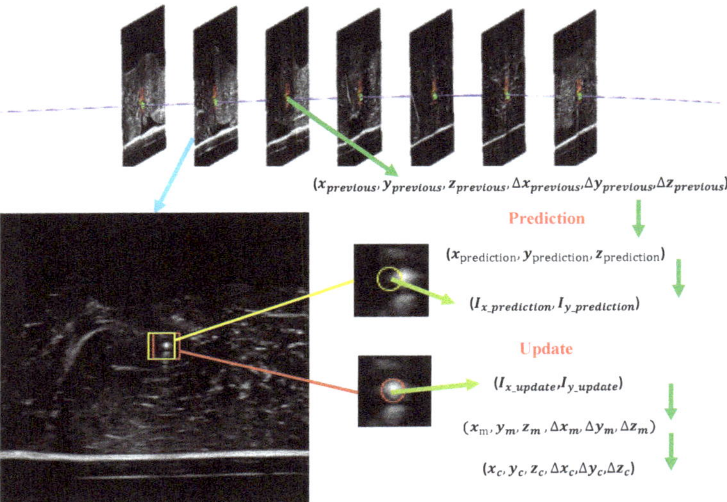

Figure 5. The two steps of the Kalman filter. As the next US image is acquired, the previous state $(x_{previous}, y_{previous}, z_{previous}, \triangle x_{previous}, \triangle y_{previous}, \triangle z_{previous})$ is used to predict the needle position $(x_{prediction}, y_{prediction}, z_{prediction},$ which then will be transformed to $(I_{x\_prediction}, I_{y\_prediction})$ in the image as the center of ROI. The yellow square is the ROI corresponding to $(I_{x\_prediction}, I_{y\_prediction})$ and the red one is the update step in KF by using the measurement data from needle detection to locate the needle with its center $(I_{x\_update}, I_{y\_update})$ as the needle position. Finally, the measurement state $(x_m, y_m, z_m, \triangle x_m, \triangle y_m, \triangle z_m)$ and the current state $(x_c, y_c, z_c, \triangle x_c, \triangle y_c, \triangle z_c)$ can be obtained.

The state prediction $\hat{t}_i$ intends to represent the prediction state of the transverse needle center position (x, y, z) in the reference frame with the change of the needle position $(\triangle x, \triangle y, \triangle z)$ at sample i according to the state t. $(\triangle x, \triangle y, \triangle z)$ are the difference between the previous needle position and the current needle position. t_i is the result from the previous iteration, which is as follows:

$$t_i = \begin{bmatrix} x_i \\ y_i \\ z_i \\ \triangle x_i \\ \triangle y_i \\ \triangle z_i \end{bmatrix} \tag{2}$$

where $\triangle x_1$ and $\triangle y_1$ are set to be 0, while $\triangle z_1$ is equal to the scan interval. The prediction equations are as follows:

$$\hat{t}_i = At_{i-1}, \tag{3}$$

$$\hat{P}_i = AP_{i-1}A^T + Q. \tag{4}$$

The measurement update equations are as follows:

$$K_i = \hat{P}_i H^T (H\hat{P}_i H^T + R)^{-1}, \tag{5}$$

$$t_i = \hat{t}_i + K_i(m_i - H\hat{t}_i),\tag{6}$$

$$P_i = (I - K_iH)\hat{P}_i,\tag{7}$$

where A, H, R, and Q are as follows:

$$A = \begin{bmatrix} 1 & 0 & 0 & 1 & 0 & 0 \\ 0 & 1 & 0 & 0 & 1 & 0 \\ 0 & 0 & 1 & 0 & 0 & 1 \\ 0 & 0 & 0 & 1 & 0 & 0 \\ 0 & 0 & 0 & 0 & 1 & 0 \\ 0 & 0 & 0 & 0 & 0 & 1 \end{bmatrix},\tag{8}$$

$$H = \begin{bmatrix} I_{6\times6} \end{bmatrix}, R = Q = 10^{-6} \times \begin{bmatrix} I_{6\times6} \end{bmatrix}.\tag{9}$$

A is the state transition matrix, H is the measurement matrix. $\hat{P}_i$ and P_i are the priori and posteriori estimate error covariance, and R and Q are the measurement error covariance and processing error covariance, respectively. K_i is the Kalman gain at sample i. m_i is the measurement state from the needle detection.

The 3D prediction position $(x_{prediction}, y_{prediction}, z_{prediction})$ is obtained from the previous state. Before needle detection in the current US image, the 3D prediction position should be transformed on the image plane frame as 2D position $(I_{x\_prediction}, I_{y\_prediction})$. After the update, the needle position $(I_{x\_update}, I_{y\_update})$ in the image will be transformed into the 3D position (x_m, y_m, z_m). Meanwhile, we can get $(\triangle x_m, \triangle y_m, \triangle z_m)$ from:

$$(\triangle x_m, \triangle y_m, \triangle z_m) = (x_m, y_m, z_m) - (x_{previous}, y_{previous}, z_{previous}).\tag{10}$$

As a result, the measurement state m_i in this frame is acquired as $(x_m, y_m, z_m, \triangle x_m, \triangle y_m, \triangle z_m)$. Through the measurement update, the current state can be obtained as $(x_c, y_c, z_c, \triangle x_c, \triangle y_c, \triangle z_c)$ from Equation (6). The relationship between these transformations will be described in the next subsection. In the previous study [26], the data from the image has only two dimensions, lacking the data from the direction along the movement of the probe, which leads to an incomplete location. Moreover, the space information is more capable to locate the needle accurately than the plane information. Therefore, 3D positions have been used in the KF for the precise location. The KF in this work is not only used for filtering, but also for predicting the next needle position in the US image. The ROI for the next iteration is centered around the needle position of the Kalman Filtering prediction, which can help to remove the outliers from the ROI.

2.3. Tip Estimation

Before tip estimation, 2D points should be transformed into 3D points based on the relationship of each frame. The relationship among the reference frame, probe frame, marker frame, and image frame are specified in Figure 6.

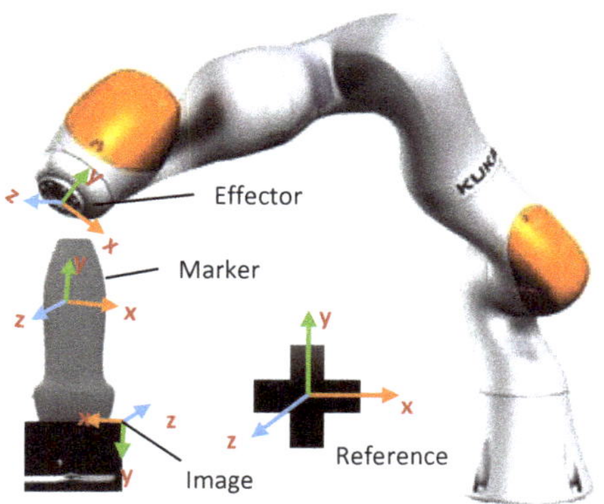

Figure 6. The relationship of the frames.

As shown in Figure 6, the image has one plane with 2 axes (axis x and axis y) and axis z is vertical to the image. Moreover, the probe has the same frame with image, except that the probe frame is designed in millimeters and the image frame is set in pixels. Equation (9) implies the transformation from image to reference:

$$Point_1 = T^{ref}_{marker} \times T^{marker}_{probe} \times T^{probe}_{image} \times Point_2 \tag{11}$$

where $Point_1$ and $Point_2$ are the points on the reference frame and image frame, respectively, T^{ref}_{marker} is the transformation from the reference frame to the marker frame, T^{marker}_{probe} is the transformation from the marker frame to the probe frame, T^{probe}_{image} is the transformation from the probe frame to the image frame. Through this transformation, the needle position in the image can be directly re-defined in the reference frame for the needle tracking and curve fitting.

The tip estimation has two steps: curve fitting and tip estimation. In this study, the third-order curve line is used for the shape construction where the equations can be written as:

$$f(x) = \sum_{k=0}^{3} a_k x^k, \tag{12}$$

$$g(x) = \sum_{k=0}^{3} b_k x^k, \tag{13}$$

where $f(x)$ and $g(x)$ are the equations to fit the line along the x, which is the axis with the same direction of the insertion. (a_0, a_1, a_2, a_3) and (b_0, b_1, b_2, b_3) are the free parameters of the needle shape model.

After sample points of the inflected needle have been obtained, the least-square curve fitting method will be used to fit these points as a cubic line. The target functions to fit the cubic line can be defined as follows:

$$F(a_0, a_1, a_2, a_3) = \min \sum_{i=1}^{n} (f(x_i) - y_i)^2 = \min \sum_{i=1}^{n} (\sum_{k=0}^{3} a_k x_i^k - y_i)^2, \tag{14}$$

$$G(b_0, b_1, b_2, b_3) = \min \sum_{i=1}^{n} (g(x_i) - z_i)^2 = \min \sum_{i=1}^{n} (\sum_{k=0}^{3} b_k x_i^k - z_i)^2, \tag{15}$$

where n is the number of the points ($n \geq 4$) and (x_i, y_i, z_i) is the position of point i. By applying the l_2 norm minimization in the two-dimensional Euclidean space, it can be formulated as:

$$\arg\min_{t \in R} \| Xa - Y \|_2,\tag{16}$$

where $a = [a_0, a_1, a_2, a_3]'$, $Y = [y_1, y_2, \ldots, y_n]'$ and X can be written as:

$$X = \begin{bmatrix} 1 & x_1 & x_1{}^2 & x_1{}^3 \\ 1 & x_2 & x_2{}^2 & x_2{}^3 \\ \vdots & \vdots & \vdots & \vdots \\ 1 & x_n & x_n{}^2 & x_n{}^3 \end{bmatrix}.\tag{17}$$

The solution can be estimated as follows:

$$a = (X^T X)^{-1} X^T Y.\tag{18}$$

From this solution, $f(x)$ and $g(x)$ can be obtained to construct needle shape. The tip position can then be estimated by the following optimum solutions based on the length of the needle:

$$\min \quad \| \int_{tail_x}^{tip_x} \sqrt{1 + f'(x)^2 + g'(x)^2} dx - L \|_2,\tag{19}$$
$$\text{s.t.} \quad tip_x > tail_x$$

where $tail_x$ is the measured value of the tail position from the optical tracker, tip_x is the expected value of the tip position in axis x, L is the length of the needle.

3. Results

3.1. Experimental Platform Setup

To verify the proposed tip tracking and shape sensing method, a robotic ultrasound system has been built, which includes a KUKA IIWA robot arm, a Wisonic ultrasound scanner, an NDI optical tracker, an NDI electromagnetic (EM) tracker, and a computer. As shown in Figure 7, the US probe is mounted on the effector of the robot arm by the gripper attached to the passive marker. The phantom or ex-vivo (like chicken breast in Figure 7) is punctured by an 18G beveled-tip needle with 200 mm long, while the needle tip is completely exposed for validation. The diameter of the needle is 15 pixels in the image. The NDI optical tracker is used to localize the marker bound with the probe, while NDI electromagnetic tracker is used to validate the tip position.

Experiments have been taken in a water tank, which provides a liquid environment for the ultrasound. And the needle is placed in water or inserted in the silica gel phantom (shown in Figure 8a), pork and chicken breast. The depth of the ultrasound is set to 4 cm. In this study, the needle is usually detected in 1 to 3 cm from the US probe. During the experiment, the robot arm automatically moves with the ultrasound probe along the direction of the needle insertion without any contact with the tissue (shown in Figure 8b). The whole scan length is at most 160 mm which depends on the scan intervals (shown in Table 1). The scan interval decreases with the increasing collected points.

Table 1. Scan lengths with different scan intervals.

Scan Interval	Scan Length	Points
1 mm	160 mm	160
2 mm	159 mm	80
3 mm	160 mm	54
4 mm	157 mm	40
5 mm	156 mm	32
6 mm	157 mm	27
7 mm	155 mm	23
8 mm	153 mm	20

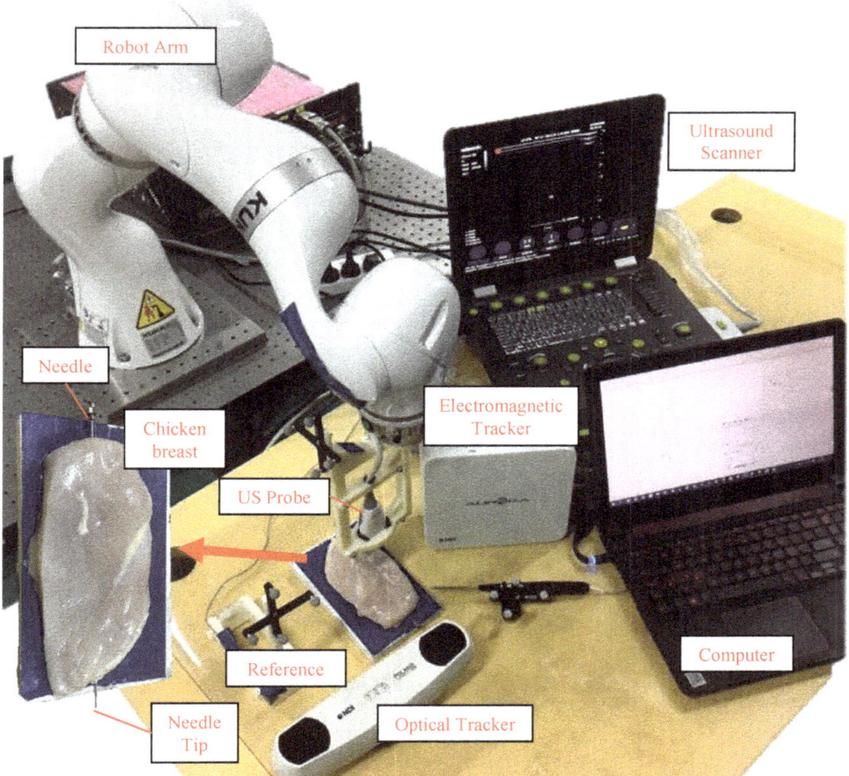

Figure 7. The devices of the experiment.

Before data collection, the US image and the marker need to be calibrated. After that, the experiment starts after the needle finished puncturing manually. The robot arm is used to move the US probe scanning along the needle. Meanwhile, pose data are collected from the optical tracker and US images from the ultrasound scanner. Finally, the tail of the needle and its tip are measured by the optical and electromagnetic sensors, respectively, for the curve fitting and the tip validation. The needle is inserted manually, imitating the real situation of percutaneous intervention.

(**a**) Phantom with inserted needle

(**b**) The probe scans without any contact

Figure 8. The phantom used in the experiment and the movement of the probe.

3.2. Tip Estimation

Four kinds of platforms have been used in the experiments: water, phantom, chicken, and pork. Each platform was tested several times. Figure 9 shows one test in chicken breast. In this case, the US probe moved along the needle in the chicken breast every 4 mm. The black square point on the left is the needle tail position and the blue line is the estimated needle shape. The green points are the detected needle and considered as the center of the needle, the blue point is the estimated tip position and the red point is measured tip position from EM. The estimation error is 0.69 mm in this test.

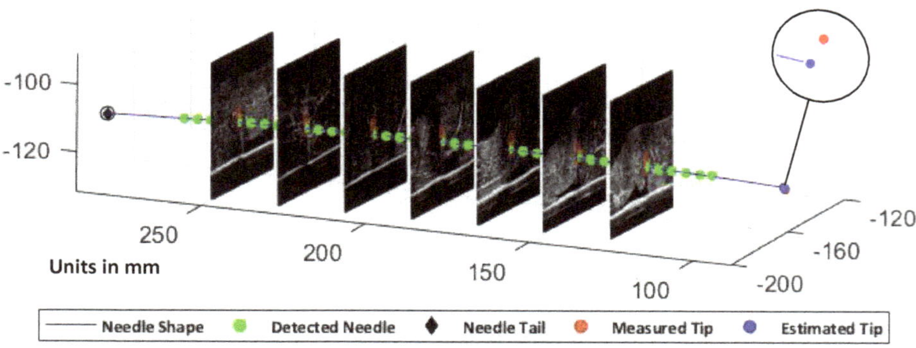

Figure 9. Experiment in chicken breast with a 4 mm scan interval.

The error of the algorithm is shown in Table 2, which suggests that the errors increase with the increase of scan intervals. Figure 10 shows the results of the experiment on four platforms. The mean errors are all under 0.4 mm with a 1 mm scan interval in the four experiments, while the error is around 1 mm with an 8 mm interval.

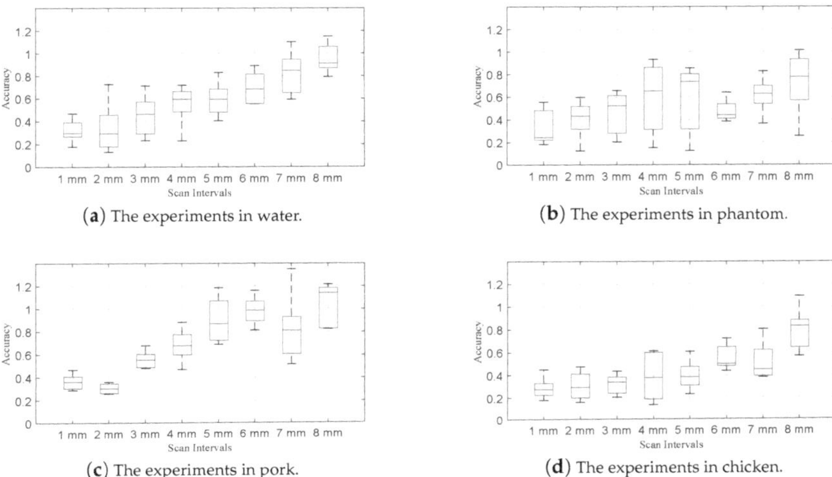

(**a**) The experiments in water.

(**b**) The experiments in phantom.

(**c**) The experiments in pork.

(**d**) The experiments in chicken.

Figure 10. The experiment with different scan intervals on the four platforms.

Table 2. The results of tip estimation (mm).

	Accuracy			
Intervals	Water	Phantom	Pork	Chicken
1 mm	0.32 ± 0.10	0.33 ± 0.15	0.37 ± 0.07	0.29 ± 0.09
2 mm	0.36 ± 0.21	0.41 ± 0.16	0.31 ± 0.05	0.31 ± 0.12
3 mm	0.45 ± 0.16	0.46 ± 0.18	0.56 ± 0.08	0.33 ± 0.09
4 mm	0.55 ± 0.16	0.59 ± 0.31	0.68 ± 0.14	0.38 ± 0.21
5 mm	0.60 ± 0.13	0.59 ± 0.29	0.90 ± 0.20	0.40 ± 0.14
6 mm	0.69 ± 0.13	0.48 ± 0.09	0.99 ± 0.13	0.55 ± 0.11
7 mm	0.83 ± 0.17	0.62 ± 0.14	0.84 ± 0.30	0.52 ± 0.17
8 mm	0.95 ± 0.11	0.73 ± 0.26	1.06 ± 0.18	0.81 ± 0.19

4. Discussion and Conclusions

Needle insertion guided by ultrasound images is widely used for percutaneous interventions. However, the needle detection due to its deflection by the inevitable factors is challenging during the needle insertion. Such factors include needle-tissue interaction, improper insertion force, physiological motions, and so on. Automatic needle detection with needle tracking in 2D transverse US images could overcome these limitations and estimate needle tip through the curve-fitting method. The target of this study is to develop a robust needle detection and tracking method with the help of ultrasound images to estimate the needle tip precisely and accurately. We used a histogram method to detect the needle in transverse US images to decrease the effects of comet tail and poor reflection. In subsequent post-processing, the needle was tracked by the Kalman filter tracking method in consecutive US images with the help of the displacement of the probe. A third-order curve fitting method has been used to estimate the needle tip. When the probe is moved by the robot arm, the scanning time is different. We assume that the time when the probe stops to collect the data is the same. The less scan interval we choose, the more collection points we can obtain and the more scanning time it takes. Therefore, the scanning time mainly depends on the number of scanning points while the accuracy lies in how short the scan interval is. In other words, the accuracy of the tip estimation can be improved by reducing the scan interval to collect more needle positions. However, this will consume more scanning time and reduce the efficiency of tracking. Inversely, fewer collecting positions would cost less time but may lead to a more possibility of the failure of shape construction and a more possibility of large error of the tip estimation. As a result, how to balance precision and scanning time is very important to make the proposed method more efficient. In our experiment, it is found that a 4 mm scan interval has an error less than 1 mm, which is a better choice to satisfy both requirements.

In the proposed method, needle shape tracking has a great contribution to the accurate localization, since the needle can be tracked precisely by Kalman filter through its prediction and update. However, needle shape tracking is heavily dependent on the scan interval, especially for a large curved needle, since Kalman filter is generally well functioned in the lineal system. As a result, if the needle is deflected during insertion, the Kalman filter would make mistakes and wrongly predict the needle position when the scan interval is large. In this study, it is found that Kalman filter would fail if the scan interval is more than 8 mm. This may due to the impact on the prediction of KF with a large deviation. Moreover, the deviation will not be eliminated even with the change of the ROI size. However, the Histogram method showed the accurate and effective detection of the needle, but it relies on the brightness of the image as the needle could not be easily detected where there are plenty of pixels with the highest intensity (which has a max value of 255). However, this condition can be controlled by the setting of the ultrasound scanner to expand the gray level of the image properly.

The proposed method still has its limitations. During the experiment, time is needed for the data collection from the US scanner and optical tracking system, and the movement of the robot arm, which is affected by the scan length and scan interval. However, it is very hard to acquire the whole position of a long needle in one scan for any medical image sensor. Therefore, in the future, it is valuable to find a method to reduce the times of needle detection in order to reduce the time for the tip estimation. Moreover, when the robot arm moves with the US probe precisely, the tissue and needle have a possibility to be deformed by the probe motion. Hence, it is necessary to make the robot arm move smoothly as well as correctly on the surface of tissue. In addition, patient motion is the biggest uncertain problem, which leads to the failure of needle insertion and detection.

In the proposed system, we use the 2D US scanner for the detection of the needle in various kinds of tissue. However, the 3D US scanner can also be used in this system. Although it has volume data and the detection method is different, the tracking method is able to be similar, as we also use the 3D positions for tracking in this study. Moreover, time is also needed for data collection and the movement of the robot arm, especially for the long needle, which is easily out of view-field of US volumes or images. Therefore, we use 2D US images in this study.

In this paper, a method for tracking a long-curved needle from the 2D transverse US images and the tip estimation is represented and demonstrated with RUS. Ultrasound is first used to detect the cross section, with the probe moving along a long-flexible needle. Needle shape tracking method combined needle detection with Kalman filter by using 3D needle positions develops an accurate location and a robust tracking procedure. Needle shape is then constructed by using the curve fitting method and its tip position is estimated based on the former result. A histogram method is introduced to detect the needle in image processing, which can improve the needle localization despite the abrupt intensity changes. This new imaging approach is proposed based on the selection of numbers of pixels with a higher gray level, which can directly remove the lower gray level to highlight the needle. Results of the experiments suggest that the detection of the needle by the histogram method and Kalman Filter has high precision with minimum error 0.13 mm with a 1 mm scan interval in the phantom experiment and maximum error 1.35 mm with a 8 mm scan interval in the pork experiment. With the increase of the scan interval, the mean error would rise. Moreover, results showed that the estimation of the tip position is less than 1 mm within 4 mm scan intervals. We suggest choosing a 4 mm scan interval to balance the precision and scanning time to maximize efficiency. In the future, we would make the experiments of how long the scan length is the best length to estimate the needle tip. The proposed method would be a great assist to surgeons to locate the needle tip when they perform percutaneous insertion procedures with a long flexible needle, such as prostate brachytherapy.

Author Contributions: Conceptualization, M.Q.-H.M.; validation, Z.L.; writing—original draft, Z.L.; writing—review and editing, S.S. and L.L.

Funding: This research was funded in part by the National Natural Science Foundation of China, grant number 61803123, and in part by the National Key R&D Program of China, grant number 2018YFB1307700, and in part by the Natural Science Foundation of Guangdong Province, China, grant number 2018A030310565.

Acknowledgments: For this kind of study, no formal ethics approval is required by the institutional ethics com. Informed consent was obtained from all individual participants included in the study.

Conflicts of Interest: The authors declare no conflict of interest.

Abbreviations

The following abbreviations are used in this manuscript:

KF	Kalman filter
US	Ultrasound
RUS	Robot ultrasound system
CT	Computed tomography
MRI	Magnetic resonance imaging
RANSAC	Random sample consensus

References

1. Orlando, N.; Snir, J.; Barker, K.; Hoover, D.; Fenster, A. Power Doppler ultrasound imaging with mechanical perturbation for improved intraoperative needle tip identification during prostate brachytherapy: A phantom study. *Proc. SPIE* **2019**, 1095131. [CrossRef]
2. Henken, K.R.; Seevinck, P.R.; Dankelman, J.; van den Dobbelsteen, J.J. Manually controlled steerable needle for MRI-guided percutaneous interventions. *Med. Biol. Eng. Comput.* **2017**. [CrossRef] [PubMed]
3. Nachabé, R.; Hendriks, B.H.W.; Schierling, R.; Hales, J.; Racadio, J.M.; Rottenberg, S.; Ruers, T.J.M.; Babic, D.; Racadio, J.M. Real-time in vivo characterization of primary liver tumors with diffuse optical spectroscopy during percutaneous needle interventions: Feasibility study in woodchucks. *Investig. Radiol.* **2015**, *50*, 443–448. [CrossRef] [PubMed]
4. Mehrjardi, M.Z.; Bagheri, S.M.; Darabi, M. Successful ultrasound-guided percutaneous embolization of renal pseudoaneurysm by autologous blood clot: Preliminary report of a new method. *J. Clin. Ultrasound* **2017**, *45*, 592–596. [CrossRef] [PubMed]
5. Jun, H.; Ahn, M.H.; Choi, I.J.; Baek, S.K.; Park, J.H.; Choi, S.O. Immediate separation of microneedle tips from base array during skin insertion for instantaneous drug delivery. *RSC Adv.* **2018**, *8*, 17786–17796. [CrossRef]
6. Park, H.; Kim, H.; Lee, S.J. Optimal Design of Needle Array for Effective Drug Delivery. *Ann. Biomed. Eng.* **2018**, *46*, 2012–2022. [CrossRef]
7. Renfrew, M.; Griswold, M.; Çavuşoğlu, M.C. Active localization and tracking of needle and target in robotic image-guided intervention systems. *Auton. Robot.* **2018**, *42*, 83–97. [CrossRef]
8. Rossa, C.; Tavakoli, M. Issues in closed-loop needle steering. *Control Eng. Pract.* **2017**, *62*, 55–69. [CrossRef]
9. Van de Berg, N.J.; Sánchez-Margallo, J.A.; van Dijke, A.P.; Langø, T.; van den Dobbelsteen, J.J. A Methodical Quantification of Needle Visibility and Echogenicity in Ultrasound Images. *Ultrasound Med. Biol.* **2019**. [CrossRef]

10. Li, R.; Xu, S.; Pritchard, W.F.; Karanian, J.W.; Krishnasamy, V.P.; Wood, B.J.; Tse, Z.T.H. AngleNav: MEMS Tracker to Facilitate CT-Guided Puncture. *Ann. Biomed. Eng.* **2018**, *46*, 452–463. [CrossRef]
11. Shellikeri, S.; Setser, R.M.; Hwang, T.J.; Srinivasan, A.; Krishnamurthy, G.; Vatsky, S.; Girard, E.; Zhu, X.; Keller, M.S.; Cahill, A.M. Real-time fluoroscopic needle guidance in the interventional radiology suite using navigational software for percutaneous bone biopsies in children. *Pediatr. Radiol.* **2017**, *47*, 963–973. [CrossRef] [PubMed]
12. Raj, S.D.; Agrons, M.M.; Woodtichartpreecha, P.; Kalambo, M.J.; Dogan, B.E.; Le-Petross, H.; Whitman, G.J. MRI-guided needle localization: Indications, tips, tricks, and review of the literature. *Breast J.* **2019**, 479–483. [CrossRef] [PubMed]
13. Ayvali, E.; Desai, J.P. Optical Flow-Based Tracking of Needles and Needle-Tip Localization Using Circular Hough Transform in Ultrasound Images. *Ann. Biomed. Eng.* **2015**, *43*, 1828–1840. [CrossRef] [PubMed]
14. Kaya, M.; Bebek, O. Needle Localization Using Gabor Filtering in 2D Ultrasound Images. In Proceedings of the 2014 IEEE International Conference on Robotics and Automation (ICRA), Hong Kong, China, 29 September 2014; pp. 4881–4886. [CrossRef]
15. Kaya, M.; Senel, E.; Ahmad, A.; Orhan, O.; Bebek, O. Real-time needle tip localization in 2D ultrasound images for robotic biopsies. In Proceedings of the IEEE International Conference on Robotics and Automation, Istanbul, Turkey, 27–31 July 2015; pp. 47–52. [CrossRef]
16. Kaya, M.; Senel, E.; Ahmad, A.; Bebek, O. Visual needle tip tracking in 2D US guided robotic interventions. *Mechatronics* **2019**, *57*, 129–139. [CrossRef]
17. Mwikirize, C.; Nosher, J.L.; Hacihaliloglu, I. Signal attenuation maps for needle enhancement and localization in 2D ultrasound. *Int. J. Comput. Assist. Radiol. Surg.* **2018**, *13*, 363–374. [CrossRef]
18. Mwikirize, C.; Nosher, J.L.; Hacihaliloglu, I. Convolution neural networks for real-time needle detection and localization in 2D ultrasound. *Int. J. Comput. Assist. Radiol. Surg.* **2018**, *13*, 647–657. [CrossRef]
19. Xu, F.; Gao, D.; Wang, S.; Zhanwen, A. MLESAC Based Localization of Needle Insertion Using 2D Ultrasound Images. *J. Phys. Conf. Ser.* **2018**, *1004*. [CrossRef]
20. Yue, Z.; Liebgott, H.; Cachard, C. Tracking biopsy needle using Kalman filter and RANSAC algorithm with 3D ultrasound. In Proceedings of the Acoustics 2012, Nantes, France, 23–27 April 2012; pp. 231–236,
21. Chatelain, P.; Krupa, A.; Navab, N. 3D ultrasound-guided robotic steering of a flexible needle via visual servoing. In Proceedings of the IEEE International Conference on Robotics and Automation, Seattle, WA, USA, 26–30 May 2015; pp. 2250–2255. [CrossRef]
22. Arif, M.; Moelker, A.; van Walsum, T. Automatic needle detection and real-time Bi-planar needle visualization during 3D ultrasound scanning of the liver. *Med. Image Anal.* **2019**, *53*, 104–110. [CrossRef]
23. Younes, H.; Voros, S.; Troccaz, J. Automatic needle localization in 3D ultrasound images for brachytherapy. In Proceedings of the 2018 IEEE 15th International Symposium on Biomedical Imaging (ISBI 2018), Washington, DC, USA, 4–7 April 2018; pp. 1203–1207. [CrossRef]
24. Vrooijink, G.J.; Abayazid, M.; Misra, S. Real-time three-dimensional flexible needle tracking using two-dimensional ultrasound. In Proceedings of the IEEE International Conference on Robotics and Automation, Karlsruhe, Germany, 6–10 May 2013; pp. 1688–1693. [CrossRef]
25. Waine, M.; Rossa, C.; Sloboda, R.; Usmani, N.; Tavakoli, M. 3D shape visualization of curved needles in tissue from 2D ultrasound images using RANSAC. In Proceedings of the IEEE International Conference on Robotics and Automation, Seattle, WA, USA, 26–30 May 2015; pp. 4723–4728. [CrossRef]
26. Waine, M.; Rossa, C.; Sloboda, R.; Usmani, N.; Tavakoli, M. Needle Tracking and Deflection Prediction for Robot-Assisted Needle Insertion Using 2D Ultrasound Images. *J. Med. Robot. Res.* **2016**, *01*, 1640001. [CrossRef]
27. Waine, M.; Rossa, C.; Sloboda, R.; Usmani, N.; Tavakoli, M. 3D Needle Shape Estimation in TRUS-Guided Prostate Brachytherapy Using 2D Ultrasound Images. *IEEE J. Biomed. Health Inform.* **2015**, *2194*, 1–11. [CrossRef]
28. Priester, A.M.; Natarajan, S.; Culjat, M. Robotic ultrasound systems in medicine. *IEEE Trans. Ultrason. Ferroelectr. Freq. Control* **2013**, *60*, 507–523. [CrossRef] [PubMed]

29. Mignon, P.; Poignet, P.; Troccaz, J. Automatic Robotic Steering of Flexible Needles from 3D Ultrasound Images in Phantoms and Ex Vivo Biological Tissue. *Ann. Biomed. Eng.* **2018**, *46*, 1385–1396. [CrossRef] [PubMed]

30. Mignon, P.; Poignet, P.; Troccaz, J. Beveled-tip needle-steering using 3D ultrasound, mechanical-based Kalman filter and curvilinear ROI prediction. In Proceedings of the 2016 14th International Conference on Control, Automation, Robotics and Vision (ICARCV), Phuket, Thailand, 13–15 November 2016; pp. 1–6. [CrossRef]

Article

Scanning Electron Microscopy Analysis and Energy Dispersion X-ray Microanalysis to Evaluate the Effects of Decontamination Chemicals and Heat Sterilization on Implant Surgical Drills: Zirconia vs. Steel

Antonio Scarano [1,2,3,*], Sammy Noumbissi [2,4], Saurabh Gupta [2,5], Francesco Inchingolo [6], Pierbiagio Stilla [1] and Felice Lorusso [7]

[1] Department of Medical, Oral and Biotechnological Sciences and CeSI-Me, University of Chieti-Pescara, 3166100 Chieti, Italy
[2] Zirconia Implant Research Group (Z.I.R.G.), International Academy of Ceramic Implantology, Silver Spring, MD 20901, USA
[3] Department of Oral Implantology, Dental Research Division, College Ingà, UNINGÁ, Cachoeiro de Itapemirim 29312, Brazil
[4] International Academy of Ceramic Implantology, Zirconia Implant Research Group (Z.I.R.G.), University of Chieti-Pescara, 3166100 Chieti, Italy
[5] Private Practice Dentistry, Bangalore 560001, India
[6] Department of Interdisciplinary Medicine, University of Bari "Aldo Moro", 70121 Bari, Italy
[7] Department of Medical, Oral and Biotechnological Sciences, University of Chieti-Pescara, Via dei Vestini, 3166100 Chieti, Italy
* Correspondence: ascarano@unich.it

Received: 3 June 2019; Accepted: 10 July 2019; Published: 16 July 2019

Featured Application: The control of the effect of drills wear and corrosion related to clinical use represents a critical factor in implantology, because it is strictly related to the bone cutting precision and heat generation during site preparation.

Abstract: Background: Drills are an indispensable tool for dental implant surgery. Today, there are ceramic zirconium dioxide and metal alloy drills available. Osteotomy drills are critical instruments since they come in contact with blood and saliva. Furthermore, they are reusable and should be cleaned and sterilized between uses. Depending on the material, sterilizing agents and protocols can alter the surface and sharpness of implant drills. The hypothesis is that cleaning and sterilization procedures can affect the surface structure of the drills and consequently reduce their cutting efficiency. **Methods:** Eighteen zirconia ceramic drills and eighteen metal alloy drills were evaluated. Within the scope of this study, the drills were not used to prepare implant sites. They were immersed for 10 min in human blood taken from volunteer subjects and then separately exposed to 50 cycles of cleansing with 6% hydrogen peroxide, cold sterilization with glutaraldehyde 2%, and autoclave heat sterilization. Scanning Electron Microscopy (SEM) and energy dispersion X-ray (EDX) microanalysis were conducted before and after each cycle and was used to evaluate the drill surfaces for alterations. **Results:** After exposure to the cleansing agents used in this study, alterations were seen in the steel drills compared to zirconia. **Conclusions:** The chemical sterilization products used in this study cause corrosion of the metal drills and reduce their sharpness. It was observed that the cycles of steam sterilization did not affect any of the drills. Zirconia drill surfaces remained stable.

Keywords: implant drills; zirconium dioxide; sterilization; disinfection; cleaning; corrosion; implant failure

---

1. Introduction

Drill wear is a phenomenon determined by the detachment of particles during the preparation of implant sites that can interfere with the process of bone healing. Considering that all materials are subject to wear and chemical attack by disinfectants, it would be desirable to have a biocompatible drill that is not susceptible to attack by the disinfectants used in clinical practice. The healing of peri-implant bone is a complex phenomenon that involves the proliferation and differentiation of pre-osteoblasts into osteoblasts. Periosteal and endosteal activity also contributes to the production of the osteoid matrix, which is followed by mineralization and organization of the bone–implant interface [1]. The success of implant surgery depends on the bone and soft tissue healing after the process of implant site preparation [2]. One of the key factors in implant success is to keep the osteotomy process as atraumatic as possible for fast, optimal recovery and healing. One of the most important factors in minimizing trauma during osteotomy is to control heat generation with irrigation, but also by utilizing drills that will cut through bone in an efficient and minimally invasive manner [3,4]. Even after a single use, drills will show signs of wear and start to lose their sharpness and cutting efficiency [5,6]. A slightly damaged or dulled drill will not only lose its cutting efficiency, but also overheat bone, leading to bone necrosis, poor healing, and an ill-defined osteotomy with a poor initial implant stability [7,8]. Among the materials used for the manufacturing of implant drills is steel, whose structural integrity can be affected by disinfectants and cleaning agents. When coming into contact with chemical disinfectants, an oxidation process takes place on the drill, giving rise to corrosion products and the release of metal particles and ions to the peri-implant tissues, which can cause inflammation, increased osteoclastic activity, and subsequent implant failure [9,10].

The phenomenon and mechanism by which the reuse of an implant drill can be a source of metal particles or ions in implant bed preparation have not been evaluated in the literature. Metal particles and ions are known to induce osteolysis, so it is important to avoid potential sources of metal ions and particles for the long-term survival of dental implants. The reuse of drills, repeated cleansing, and sterilization causes them to lose their sharpness and trigger the corrosion process. As a result, there is a loss of cutting efficiency, thus making it possible to leave metal particles and ions in the bone. Furthermore, the loss of cutting efficiency increases the friction between the bone and drills and elevates the osteotomy temperature, leading to bone cell death and implant failure. An alternative drill material to steel is zirconium dioxide, also known as zirconia, which is a ceramic and belongs to the category of materials called inert metal oxides. They are also used as dental implant materials [11]. Zirconia is bio-inert, and has excellent corrosion resistance, superior biocompatibility, high wear resistance, and high fracture resistance values [12,13]. Repeated disinfection and sterilization cycles can lead to undesirable changes in the physical properties of the instruments, such as a loss of structure, sharpness, and corrosion [5]. The purpose of this study is to analyze the effects of chemical disinfection and autoclave sterilization on zirconium oxide ceramic drills vs. steel drills by scanning electron microscopy and energy dispersion X-ray microanalysis. The null hypothesis of this study is that there will be no difference in surface alterations between zirconia and steel drills as a result of disinfection and sterilization procedures.

2. Materials and Methods

In this study, a total of 18 drills of zirconia (Dental Tech, Misinto, Milan) and 18 steel drills were used (Dental Tech, Misinto, Milan) (Figure 1).

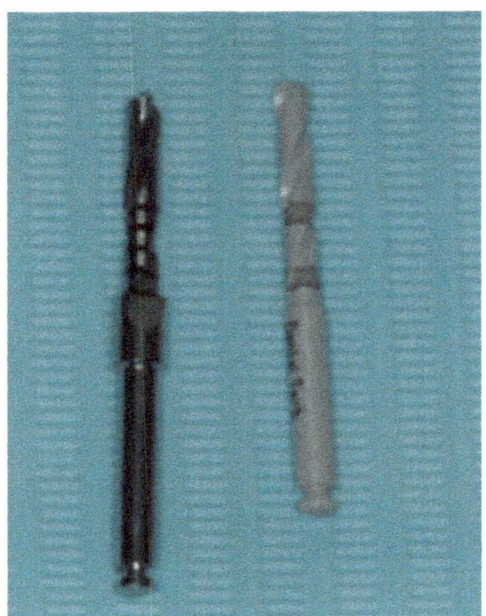

Figure 1. Macroscopic aspect of the steel drill (left) and zirconia drill (right) used in the present study.

Only 2 mm diameter helical conical drills were evaluated in order to analyze their response to disinfecting and cleansing agents, as well as autoclave sterilization. The steel drills presented the following chemical composition: 0.2% sulfur, 0.2% carbon, 0.6% silicon, 0.8% nickel, 1.2% molybdenum, 1.6% manganese, 16% chromium, 22% carbon, and balancing iron. The chemical composition shows the typical characteristics of a stainless-steel alloy. The drills studied were all externally irrigated and had three black laser markings with the function of guiding the surgeon in controlling the depth during osteotomy. In this investigation, chemical agents such as glutaraldehyde 2% (Dimexid, Amedics Ferrara, Italy) and hydrogen peroxide 6% (Sanibios, Noda, Balerna- Switzerland), were taken into consideration, and the effect of sterilization cycles was also observed. The drills were not subjected to the preparation of implant sites, but were immersed for 10 minutes in human blood taken from voluntary subjects and then underwent cycles of immersion in disinfectants without being subjected to autoclaving (approved by Inter Institutional Ethics Committee of UNINGÁ, No. 72105917.5.0000.5220). Before immersion, decontamination, or sterilization, the drills were rinsed in running water and scrubbed with a toothbrush made of nylon bristles in order to remove gross foreign material (e.g., organic material, clot). A total of 18 steel drills and 18 zirconia drills were used and allotted to 6 groups of 6 drills, with 6 in steel and 6 in zirconia for each group. Then, implant drills were assessed before and after the disinfection by Scanning Electron Microscopy (SEM). A disc of carborundum was used to separate the blades from the grips of the drills to be observed by SEM. Ten flat areas of 200 μm to 150 μm in diameter were evaluated for each drill and an image in JPEG format was created (Figure 2).

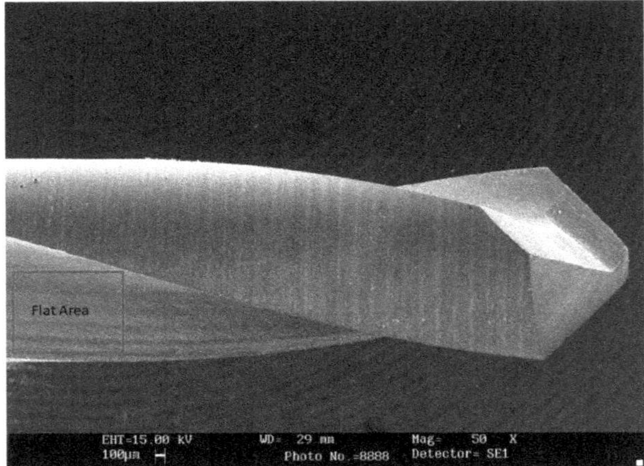

Figure 2. The flat area evaluated for each drill.

The drills were divided into the following groups:

Group 1: New non-sterilized Steel drill; Steel drills subjected to 50 cycles of 20 min sterilization immersion in glutaraldehyde;

Group 2: Steel drills subjected to 50 immersion cycles of 20 min immersion cleansing in hydrogen peroxide;

Group 3: Steel drills subjected to 50 cycles of heat sterilization in an autoclave at 134° C for 35 min at a pressure of 1.1 bar; New non-sterilized Zirconia drills;

Group 4: Zirconia drills subjected to 50 cycles of 20 min immersion sterilization in glutaraldehyde;

Group 5: Zirconia drills subjected to 50 immersion cycles of 20 min for cleansing in hydrogen peroxide;

Group 6: Zirconia drills subjected to 50 cycles of heat sterilization in an autoclave at 134° C for 35 min at a pressure of 1.1 bar (A-17 PLUS, ANTHOS, Imola, Italy).

All the solutions were poured into a plastic container in order to avoid interference between the liquid and metal. In accordance with the Italian legislation DMS of 28/9/90, the drills were immersed for 20 min in a liquid chemical sterilant. After 50 cycles of decontamination and cold sterilization immersion, the drills were rinsed in running water and then in demineralized water. They were then observed by a transmission electron microscope. The method of analysis followed the methodology proposed by Scarano et al. [5].

Previously calibrated examiners compared the micrographs before and after 50 cycles of disinfection and sterilization, respectively, with those obtained for the control group and assessed the percentage of surface covered by damage.

2.1. SEM Observations

In order to evaluate changes in the surface topography, we used a scanning electron microscope (SEM, JSM-6480LV; Jeol, Tokyo, Japan), with a solid-state backscattered detector operated at a 20 kV accelerating voltage. Each drill was attached to an aluminum stub with sticky conductive carbon tape. Images were taken in both secondary and backscattered electrons. The drills were mounted on aluminum stubs and gold-coated in Emitech K550 (Emitech Ltd., Ashford, UK). Pictures were recorded for each specimen to characterize the surface and presence of corrosion area. The pictures were processed with ImageJ (ImageJ, U.S. National Institute of Health, Bethesda, MD, USA).

2.2. Energy Dispersive X-ray Spectroscopy (EDX)

To detect possible drill alterations, the drills were examined before and after sterilization with a scanning electron microscope/energy dispersion X-ray microanalysis (SEM/EDX) using scanning electron microscopy (SEM, JSM-6480LV; Jeol, Tokyo, Japan) with the EDS QUANTAX-200 probe (Bruker Nano GmbH, Berlin, Germany).

2.3. Statistical Evaluation

Free software available on the site http://clincalc.com/stats/samplesize.aspx, was used to determine the number of drills needed to achieve statistical significance for quantitative analyses of drill surface damage. A calculation model was adopted for dichotomous variables (yes/no effect) by applying the effect incidence designed to caution the reasons, with values of 10% for zirconia drill and 95% for steel drill, and the alpha was set at 0.05 and power at 70%. The optimal number of specimens for analysis was six drills per group.

A normal data distribution was evaluated by the Shapiro–Wilk test and the differences between the groups were analyzed by one-way analysis of variance (ANOVA) followed by a Tukey post-hoc test. A p-value < 0.05 was considered statistically significant. Data processing and statistical analysis were performed by Excel origin and SPSS software (SPS, Bologna, Italy).

3. Results

Scanning Electron Microscopy Analysis and Energy Dispersion X-ray Microanalysis.

3.1. Steel Drills

New Non-Sterilized Drill

The spectroscopic analysis showed that the flat area of implant drills was composed of stainless-steel alloys with a high content of iron, nickel, and chromium (Figure 3).

Group 1

After 50 cycles of decontamination, the drills immersed in the glutaraldehyde showed traces of corrosion with deposits on the bottom of the container and traces of a rust color. The drills appeared to be macroscopically damaged and the laser depth markings were less distinctive, but still identifiable. The areas of corrosion were mostly present especially near the laser markings (Table 1). In total, $21 \pm 3\%$ of the drill surface in this group was covered by damage (Figure 4). None of the drills had surface damage over 30% of the total surface. The spectroscopic analysis of the drill surface showed a reduction in iron, nickel, and chromium and a comparable increase in oxygen (Figure 5).

Group 2

The drills immersed in hydrogen peroxide appeared to be macroscopically damaged, but the damaged surface areas were smaller and the laser depth marks remained clearly visible. A total of $12 \pm 1\%$ of the drill surface in the hydrogen peroxide group was covered by damage. No drill in this group had damage above 20% of the surface (Figure 6). Additionally, in this group, the spectroscopic analysis of the drill surface showed a decrease in iron levels and a comparable increase in oxygen.

Group 3

The steel drills that were subjected to 50 cycles of heat sterilization by steam neither caused areas of corrosion nor altered the depth notches (Table 1) (Figure 7). The spectroscopic analysis of the drill surface showed a smaller increase in oxygen and a decrease in the levels of iron, nickel, and chromium (Figure 8).

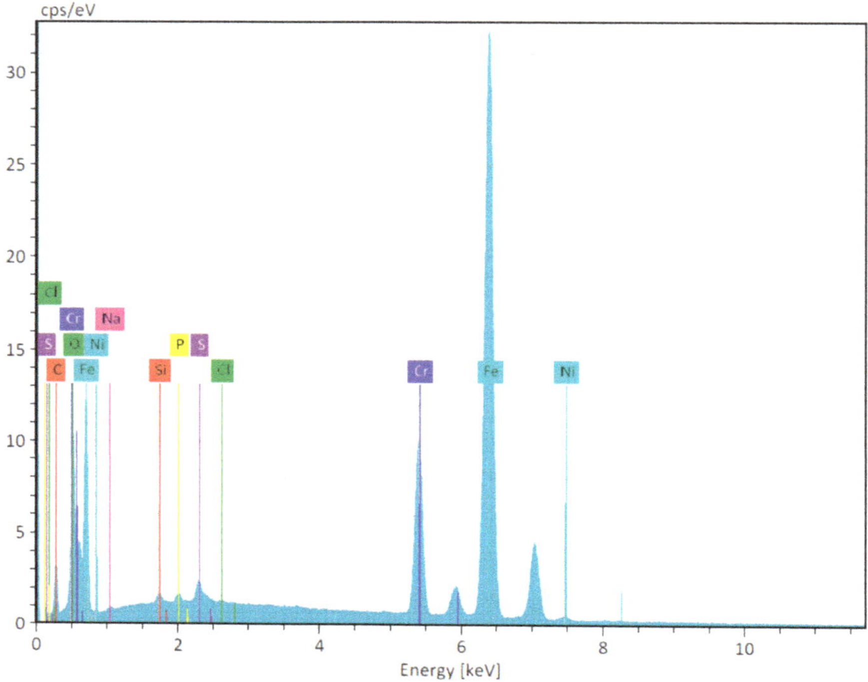

Figure 3. Spectra showing elemental peaks of the new drill surface.

Figure 4. Group 1. Aspects of the steel drill after 50 disinfection cycles with glutaraldehyde, where zones of structural alteration can be observed.

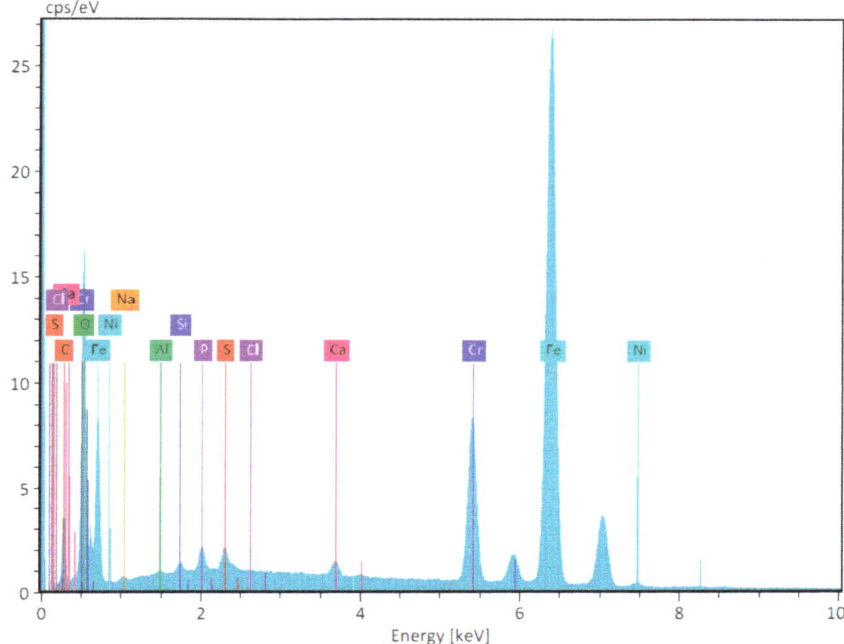

Figure 5. Group 1. The spectroscopic analysis of the drill surface after 50 disinfection cycles with glutaraldehyde showed a decrease in oxygen, iron, nickel, and chromium.

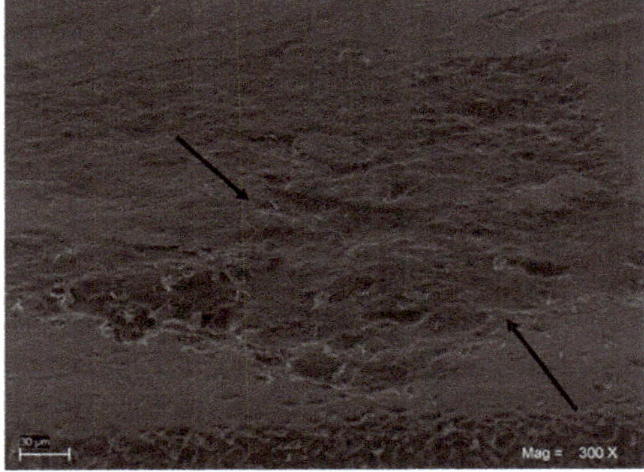

Figure 6. Group 2. After 50 disinfection cycles with hydrogen peroxide, zones of damage by the corrosive process were observed (arrows).

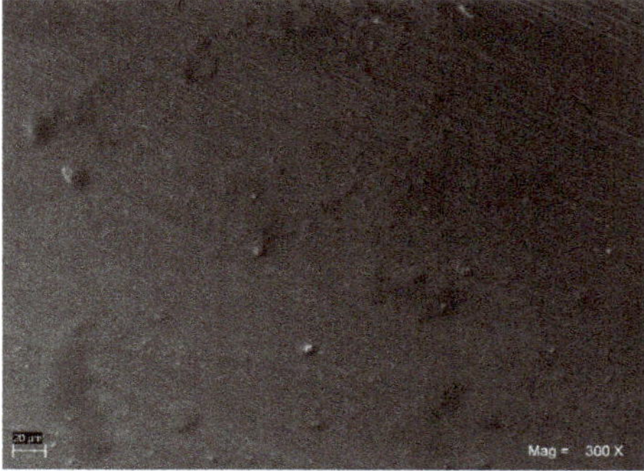

Figure 7. Group 3. After 50 steam autoclave cycles, no areas of corrosion were observed.

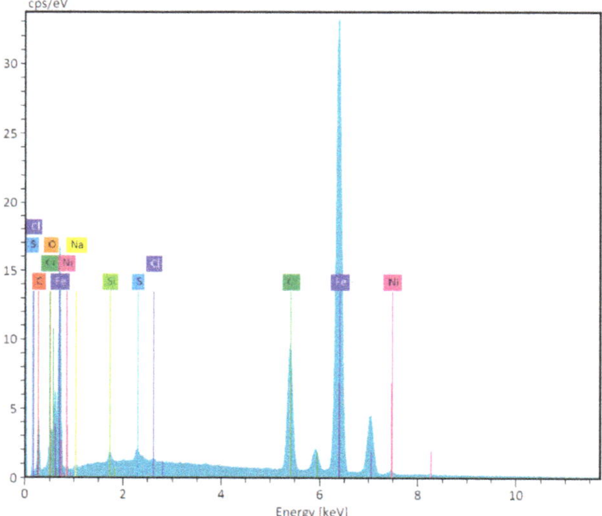

Figure 8. Group 3. The spectroscopic analysis of the drill surface after 50 steam autoclave cycles, where no changes in iron, nickel, and chromium levels and a smaller increase in oxygen where observed.

3.2. Zirconia Drills

New Non-Sterilized Drill

The spectroscopic analysis of the new unsterilized zirconia drill surface showed that the flat areas of implant drills were mainly composed of zirconia and carbon, oxygen, and aluminum (Figure 9).

Group 4

The zirconia ceramic drills observed after 50 cycles of immersion in the glutaraldehyde had the same appearance as the new non-sterilized drills and did not have residues of organic material (Figure 10). The spectroscopic analysis of the drill surface showed no change in the surface composition.

Group 5

The zirconia ceramic drills observed after 50 cycles of immersion in the hydrogen peroxide had the same appearance as the new non-sterilized drills and did not have residues of organic material (Figure 11). The spectroscopic analysis of the drill surface showed no change in the surface composition.

Group 6

No damage was detected after 50 cycles of heat sterilization in the autoclave, and there was no corrosion or alteration of the laser markings (Figure 12). No differences were observed between the different disinfection liquid immersions (glutaraldehyde and hydrogen peroxide). The black notches that showed the depth and the numbers printed on the drill shanks were intact and showed no signs of damage.

The spectroscopic analysis of the drill surface showed no change in the surface composition.

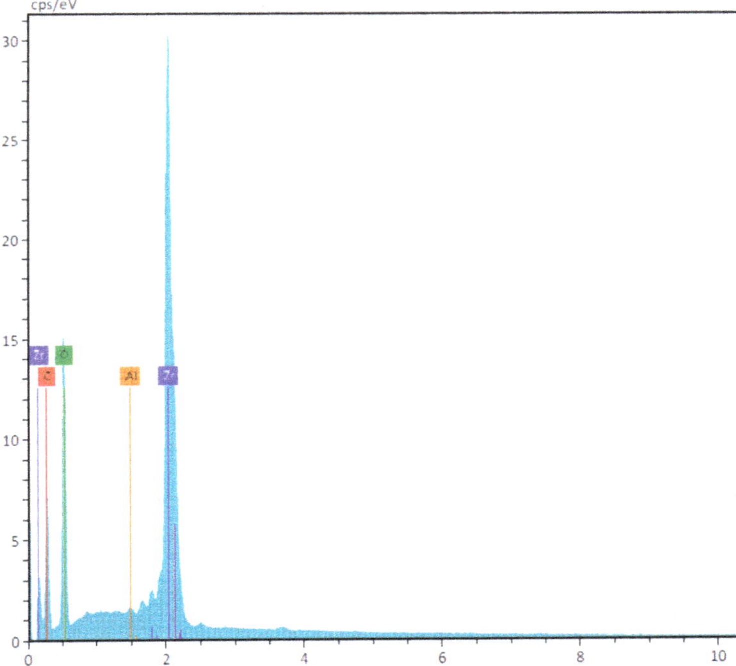

Figure 9. The spectroscopic analysis of the new zirconia drill surface showed that the flat area of implant drills was mainly composed of zirconia.

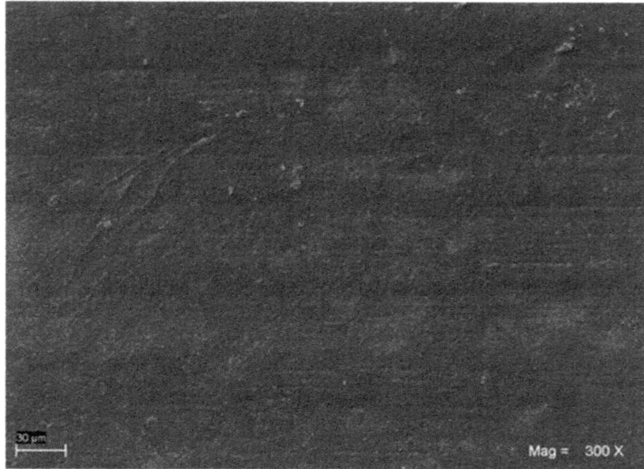

Figure 10. Group 4. The zirconia drill viewed under a scanning electron microscope after 50 disinfection cycles with glutaraldehyde. Sharp margins are free from defects.

Figure 11. Group 5. The zirconia drill after 50 disinfection cycles with hydrogen peroxide, where no damage was observed on the cutting edges or deposits of other substances.

Figure 12. Group 6. After 50 steam autoclave cycles, no areas of corrosion were observed.

Table 1. Summary of the drills surface area damaged by the testing processes. The disinfection processes generated significant surface corrosion compared to the zirconia drills ($p < 0.01$). The autoclave sterilization did not cause any alteration to the different drill surfaces.

	STEEL DRILLS			ZIRCONIA DRILLS		
	Group I (6 drills)	**Group II** (6 drills)	**Group III** (6 drills)	**Group IV** (6 drills)	**Group V** (6 drills)	**Group VI** (6 drills)
Mean	21% **	12% **	No Damage **	No Damage **	No Damage **	No Damage **
SD	±3%	±2%	–	–	–	–

$** p < 0.01.$

3.3. Statistical Analysis

The SEM results showed the drill surface covered by damage and corrosion percentages for the steel drill. A statistically significant difference was found in the percentage of drill surface covered by damage and corrosion on the zirconia drill compared to the steel implant drills ($p = 0.000034$) (Figure 13).

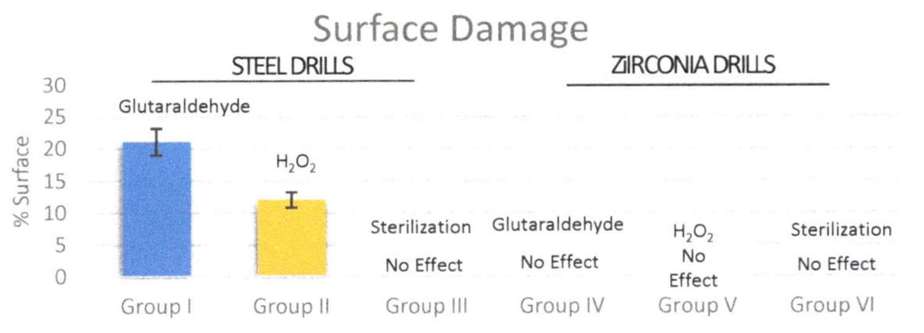

Figure 13. Differences in surface alteration occurred in the different study groups.

4. Discussion

In clinical practice, it is important to avoid cross-infection and it is for this reason that the drills should be treated with disinfectant chemicals, thoroughly rinsed, and submitted to autoclave heat sterilization.

In this study, we used two different disinfection chemicals, namely glutaraldehyde and hydrogen peroxide, which both provide high-level disinfection. The outcomes of the present research showed that zirconia drills are more resistant to the corrosive action of disinfecting chemicals. In fact, a greater difference was found in the percentages of drill covered by surface damage. SEM analysis demonstrated that repeated autoclave sterilization cycles had no effect on the zirconia and steel drills.

The purpose of the present investigation was to study the influence of disinfection and sterilization on the implant drill surfaces. The authors hypothesized that zirconia drills may offer a greater resistance to the action of disinfectants commonly used in clinical practice.

The contact with contaminated surfaces produces destructive hydroxyl free radicals that can attack DNA, membrane lipids, and other essential cell components. Glutaraldehyde is a saturated dialdehyde that has gained wide acceptance as a high-level disinfectant and chemical sterilant [14]. Aqueous solutions of glutaraldehyde are "activated" by using alkalinating agents to bring the pH to values between 7.5 and 8.5, which makes the solution sporicidal [15,16]. Glutaraldehyde (GAA) solutions are used for the sterilization of medical devices, such as endoscopes and fibroscopes. Hydrogen peroxide has good bactericidal, sporicidal, viricidal, and fungicidal properties [17]. However, both glutaraldehyde and hydrogen peroxide are corrosive to steel instruments.

Implant drills are reusable medical devices and implant bed preparation with drills is an invasive procedure involving contact of the drill with blood, bone, and other biological fluids. The major risk of all such procedures is the introduction of drill particles and transfer of pathogens, potentially leading to infection [18]. The main objective of disinfecting or sterilizing drills is to eliminate the transmission of pathogens by means of contaminated medical and surgical devices (e.g., HIV, HCV, *Mycobacterium tuberculosis*, and encephalopathies) [19,20]. Today, the disinfection of medical devices such as implant drills is very important for preventing the emergence of transmissible spongiform, such as the encephalopathies (TSEs), which is a prion protein less susceptible to denaturation by heat and is responsible for disease such as variant Creutzfeldt–Jakob [21]. Therefore, the predicable cleaning of implant drills is believed to be a key procedure for reducing the risks of onward transmission of infectious diseases and pathogens such as human immunodeficiency virus (HIV), hepatitis B and C, gram-negative and gram-positive bacteria, fungi, and preventing the risk of mycobacteria transmission. In addition to microbial inactivation, the removal of organic or inorganic debris is crucial as it may compromise subsequent disinfection or sterilization processes [22]. The cleaning of reusable implant drills is also important in order to ensure drill longevity and efficiency in cutting bone, also making the removal of organic and chemical residues important. The effect of liquid chemical sterilant depends on cleaning to eliminate organic and inorganic material, an optimal pH, concentration, contact time, and temperature are all paramount. During chemical disinfection, it is important that implant drills are not mixed with other drills, or other instruments of different material composition in order to avoid chemical corrosion. In this study, once chemical disinfection was completed, we proceeded to heat sterilization at 134 °C (273 °F) for 35 min.

It is known that the effect of disinfectants and sterilization can influence the cutting efficiency of drills [5,23]. The material that constitutes an osteotomy drill must also have a good resistance to corrosion, which is one of the factors responsible for the loss of cutting efficiency. The drills used for the preparation of implant sites must undergo decontamination and sterilization procedures before being reused on another patient. After every use, reusable medical devices must be rinsed, cleaned, and immediately immersed in an approved medical grade chemical disinfectant with a recognized effectiveness against HIV in accordance with Italian legislation art. 2 Ministerial Decree 28.09.90 [24]. The Ministerial Decree (DMS) imposes a set of rules for the protection of professionals from a broad range of infections in public and private health and care facilities. The objectives of these precautions are to prevent the transmission of pathogenic organisms through blood and biological fluids. The cleansing occurs as a result of the chemical actions of a detergent that must be used at recommended concentrations and contact times. It is necessary to frequently renew the solution in order to avoid the accumulation of debris and contamination, which both decrease the detergent efficacy.

Implant surgical procedures expose patients, surgeons, and staff to potential cross-infections if carried out without a series of cleaning and sterilization procedures aimed at preventing the spread of pathogens. Bone drills are classified as critical surgical objects according to the classification proposed by E.H. Spaulding [25] and from the DMS dated 28/9/90 as instruments that come in contact with bleeding tissues and body fluids. For this reason, implantology drills must also be subjected to decontamination and cleaning procedures using chemical substances before proceeding to heat sterilization with saturated steam in an autoclave. All these procedures and protocols can alter the cutting efficiency, as well as deteriorate the depth marks printed on the drills, thereby altering the references that the surgeon has during surgery to check the working depth of the osteotomy. The reduction in cutting efficiency has considerable repercussions for the preparation of the implant site because a loss of sharpness and cutting efficiency of the drill results in a large part of the cutting energy being transformed into heat. The consequences manifest themselves as poor organization of the clot, delayed healing, and poor quality of the tissues that form at the bone–implant interface [18,26]. As reported previously in another study [18], implant osteotomy drills can be reused up to 50 times since their loss of sharpness is directly proportional to the number of osteotomies made.

Chemical detergents can cause metal corrosion with alteration of the cutting edge of the drill blades, which is a process that can worsen when the drills are submitted to saturated steam sterilization. In the literature, there are several articles that have studied the degree of wear of osteotomy drills in relation to their reuse [26,27]. On the other hand, there is no published work investigating the problem of chemical decontamination or metal release in the bone from surgical drills. In the present work, we chose to separately study the effect on bone drills of the two different chemical agents and the effect of autoclave sterilization in order to understand which of these aspects of bone instrument care is the most critical and potentially damaging to the drills.

We observed that all the disinfectants used were found to be potentially harmful to steel drills, causing corrosion zones that worsened after each decontamination cycle, while the zirconia drills did not show any structural damage, and the depth marks were preserved and remained clearly visible. No damage was observed for the steel and zirconia drills when they were only subjected to autoclave sterilization cycles. The results of the present study provide practical and valuable information that can help conserve the cutting efficiency of the drills by preventing metal release and overheating of the bone, both of which can increase the chances of implant failure [28].

Identifying factors and mechanisms with regards to implant failures should not be limited to the implants, but should also include the host and the drills. Drills are one of the main components of an implant procedure and play an important role in the success of an implant procedure [29,30]. Therefore, it is of primary importance to identify critical factors and develop appropriate therapies and prevention strategies. It is extremely useful to know the histopathology (macroscopic and microscopic aspects of the disease), the pathogenesis (source of the disease), and the physiopathology (mechanism of the disease) of the complications of implant failures, as well as changes in the structure of the bone drills during the processes of decontamination, cleaning, and sterilization.

From the results observed during this study, it is clear that the disinfectants used in clinical practice are potentially harmful to steel drills by contributing to corrosion of their surfaces and the loss of cutting efficiency.

5. Conclusions

The present study demonstrates that repeated immersions of steel drills in disinfecting chemical agents result in corrosion on the surface of steel drills. Conversely, these chemicals do not have any effect on zirconia ceramic drills, proving their inertness and structural stability, even in the harshest environment. Autoclave sterilization cycles have no effect on any of the drills. Cleansing, sterilization, and maintenance of the drills are crucially important for patients' health and protection, but also for the drills' long-term durability and performance. It can be concluded that zirconia ceramic drills offer a surface that is not affected by disinfecting liquids. The fact that zirconia is immune to corrosion attack eliminates the possibility of the release of microparticles and ions in the peri-implant tissue during osteotomy, thus minimizing the chance of aseptic osteolysis [31].

Author Contributions: Conceptualization, A.S.; data curation, A.S.; formal analysis, S.N. and S.G.; investigation, A.S.; methodology, A.S.; project administration, A.S.; software, F.L. and P.S.; supervision, P.S.; F.I.; validation, A.S., S.N., F.I., and S.G.; visualization, P.S. and F.L.; writing—original draft, A.S.; writing—review and editing, S.N.

Funding: This research received no external funding.

Conflicts of Interest: The authors declare no conflict of interest.

References

1. Scarano, A.; Piattelli, A.; Quaranta, A.; Lorusso, F. Bone Response to Two Dental Implants with Different Sandblasted/Acid-Etched Implant Surfaces: A Histological and Histomorphometrical Study in Rabbits. *BioMed Res. Int.* **2017**, *2017*, 8724951. [CrossRef] [PubMed]
2. Eriksson, A.; Albrektsson, T. Temperature threshold levels for heat-induced bone tissue injury: A vital-microscopic study in the rabbit. *J. Prosthet. Dent.* **1983**, *50*, 101–107. [CrossRef]
3. Scarano, A.; Carinci, F.; Lorusso, F.; Festa, F.; Bevilacqua, L.; Santos de Oliveira, P.; Maglione, M. Ultrasonic vs Drill Implant Site Preparation: Post-Operative Pain Measurement through VAS, Swelling and Crestal Bone Remodeling: A Randomized Clinical Study. *Materials* **2018**, *11*, 2516. [CrossRef] [PubMed]
4. Eriksson, R.; Adell, R. Temperatures during drilling for the placement of implants using the osseointegration technique. *J. Oral Maxillofac. Surg.* **1986**, *44*, 4–7. [CrossRef]
5. Scarano, A.; Di Carlo, F.; Piattelli, A. Effect of sterilization and cleansing on implantology drills: Zirconia vs steel. *Ital. Oral Surg.* **2008**, *3*, 61–72.
6. Hochscheidt, C.J.; Shimizu, R.H.; Andrighetto, A.R.; Pierezan, R.; Thomé, G.; Salatti, R. Comparative Analysis of Cutting Efficiency and Surface Maintenance between Different Types of Implant Drills: An In Vitro Study. *Implant Dent.* **2017**, *26*, 723–729. [CrossRef] [PubMed]
7. Hein, C.; Inceoglu, S.; Juma, D.; Zuckerman, L. Heat Generation during Bone Drilling: A Comparison between Industrial and Orthopaedic Drill Bits. *J. Orthop. Trauma* **2017**, *31*, e55–e59. [CrossRef] [PubMed]
8. Louropoulou, A.; Slot, D.E.; Van der Weijden, F.A. Titanium surface alterations following the use of different mechanical instruments: A systematic review. *Clin. Oral Implant. Res.* **2012**, *23*, 643–658. [CrossRef]
9. Noumbissi, S.; Scarano, A.; Gupta, S. A Literature Review Study on Atomic Ions Dissolution of Titanium and Its Alloys in Implant Dentistry. *Materials* **2019**, *12*, 368. [CrossRef]
10. Dubruille, J.H.; Viguier, E.; Le Naour, G.; Dubruille, M.T.; Auriol, M.; Le Charpentier, Y. Evaluation of combinations of titanium, zirconia, and alumina implants with 2 bone fillers in the dog. *Int. J. Oral Maxillofac. Implant.* **1999**, *14*, 271–277.
11. Scarano, A.; Di Carlo, F.; Quaranta, M.; Piattelli, A. Bone Response to Zirconia Ceramic Implants: An Experimental Study in Rabbits. *J. Oral Implant.* **2003**, *29*, 8–12. [CrossRef]
12. Jackson, M.C. Restoration of posterior implants using a new ceramic material. *J. Dent. Technol.* **1999**, *16*, 19–22. [PubMed]
13. Spinelli, M.; Maccauro, G.; Graci, C.; Cittadini, A.; Magnani, G.; Sangiorgi, S.; Del Bravo, V.; Manicone, P.; Raffaelli, L.; Muratori, F.; et al. Zirconia Toughened Alumina (ZTA) Powders: Ultrastructural and Histological Analysis. *Int. J. Immunopathol. Pharmacol.* **2011**, *24*, 153–156. [CrossRef] [PubMed]
14. Lin, W.; Niu, B.; Yi, J.; Deng, Z.; Song, J.; Chen, Q. Toxicity and Metal Corrosion of Glutaraldehyde-Didecyldimethylammonium Bromide as a Disinfectant Agent. *BioMed Res. Int.* **2018**, *2018*, 9814209. [CrossRef] [PubMed]
15. Urayama, S.; Kozarek, R.A.; Sumida, S.; Raltz, S.; Merriam, L.; Pethigal, P. Mycobacteria and glutaraldehyde: Is high-level disinfection of endoscopes possible? *Gastrointest. Endosc.* **1996**, *43*, 451–456. [CrossRef]
16. Chenjiao, W.; HongYan, Z.; Qing, G.; Xiaoqi, Z.; Liying, G.; Ying, F. In-Use Evaluation of Peracetic Acid for High-Level Disinfection of Endoscopes. *Gastroenterol. Nurs.* **2016**, *39*, 116–120. [PubMed]

17. Omidbakhsh, N.; Sattar, S.A. Broad-spectrum microbicidal activity, toxicologic assessment, and materials compatibility of a new generation of accelerated hydrogen peroxide-based environmental surface disinfectant. *Am. J. Infect. Control* **2006**, *34*, 251–257. [CrossRef]

18. Scarano, A.; Carinci, F.; Quaranta, A.; Di Iorio, D.; Assenza, B.; Piattelli, A. Effects of Bur Wear during Implant Site Preparation: An in Vitro Study. *Int. J. Immunopathol. Pharmacol.* **2007**, *20*, 23–26. [CrossRef]

19. Bagg, J.; Smith, A.J.; Hurrell, D.; McHugh, S.; Irvine, G. Pre-sterilisation cleaning of re-usable instruments in general dental practice. *Br. Dent. J.* **2007**, *202*, E22. [CrossRef]

20. Laheij, A.M.G.A.; Kistler, J.O.; Belibasakis, G.N.; Välimaa, H.; de Soet, J.J.; European Oral Microbiology Workshop (EOMW) 2011. Healthcare-associated viral and bacterial infections in dentistry. *J. Oral Microbiol.* **2012**, *4*. [CrossRef]

21. Bourgeois, D.; Dussart, C.; Saliasi, I.; Laforest, L.; Tramini, P.; Carrouel, F. Observance of Sterilization Protocol Guideline Procedures of Critical Instruments for Preventing Iatrogenic Transmission of Creutzfeldt-Jakob Disease in Dental Practice in France, 2017. *Int. J. Environ. Res. Public Health* **2018**, *15*, 853. [CrossRef] [PubMed]

22. Douet, J.Y.; Lacroux, C.; Aron, N.; Head, M.W.; Lugan, S.; Tillier, C.; Huor, A.; Cassard, H.; Arnold, M.; Beringue, V.; et al. Distribution and Quantitative Estimates of Variant Creutzfeldt-Jakob Disease Prions in Tissues of Clinical and Asymptomatic Patients. *Emerg. Infect. Dis.* **2017**, *23*, 946–956. [CrossRef] [PubMed]

23. Scarano, A.; Assenza, B.; Di Iorio, D.; Quaranta, A. Effect of sterilization and cleansing on implantology drills. *Ital. Oral Surg.* **2008**, *2*, 4–13.

24. Ministero della Sanità. *Norme di Protezione dal Contagio Professionale da HIV Nelle Strutture Sanitarie e Assistenziali Pubbliche e Private*; Ministero della Sanità: Rome, Italy, 1990.

25. Spaulding, E.H. Chemical disinfection and antisepsis in the hospital. *J. Hosp. Res.* **1972**, *9*, 5–31.

26. Tehemar, S.H. Factors affecting heat generation during implant site preparation: A review of biologic observations and future considerations. *Int. J. Oral Maxillofac. Implant.* **1999**, *14*, 127–136.

27. Rutala, W.A.; Weber, D.J. Disinfection, sterilization, and antisepsis: An overview. *Am. J. Infect. Control* **2016**, *44*, e1–e6. [CrossRef] [PubMed]

28. Scarano, A.; Piattelli, A.; Assenza, B.; Carinci, F.; Di Donato, L.; Romani, G.L.; Merla, A. Infrared thermographic evaluation of temperature modifications induced during implant site preparation with cylindrical versus conical drills. *Clin. Implant Dent. Relat. Res.* **2011**, *13*, 319–323. [CrossRef] [PubMed]

29. Sridhar, S.; Wilson, T.G.; Valderrama, P.; Watkins-Curry, P.; Chan, J.Y.; Rodrigues, D.C. In Vitro Evaluation of Titanium Exfoliation during Simulated Surgical Insertion of Dental Implants. *J. Oral Implant.* **2016**, *42*, 34–40. [CrossRef]

30. Scarano, A.; Piattelli, A.; Polimeni, A.; Di Iorio, D.; Carinci, F. Bacterial Adhesion on Commercially Pure Titanium and Anatase-Coated Titanium Healing Screws: An In Vivo Human Study. *J. Periodontol.* **2010**, *81*, 1466–1471. [CrossRef]

31. Scarano, A.; Cholakis, A.; Piattelli, A. Histologic Evaluation of Sinus Grafting Materials after Peri-implantitis–Induced Failure: A Case Series. *Int. J. Oral Maxillofac. Implant.* **2017**, *32*, e36–e75. [CrossRef]

Concept Paper

A New Concept Compliant Platform with Spatial Mobility and Remote Actuation

Nicola Pio Belfiore

Department of Engineering, University of Roma Tre, via della Vasca Navale 79, 00146 Rome, Italy; nicolapio.belfiore@uniroma3.it; Tel.: +39-06-5733-3316

Received: 27 June 2019; Accepted: 19 September 2019; Published: 21 September 2019

Abstract: This paper presents a new tendon-driven platform with spatial mobility. The system can be obtained as a monolithic structure, and its motion is based on the concept of selective compliance. The latter contributes also to optimizing the use of the material by avoiding parasitic deformations. The presented platform makes use of lumped compliance with three different kinds of elastic joints. An analysis of the platform mobility based on finite element analysis is provided together with an assembly mode analysis of the equivalent pseudo-rigid body mechanism. Surgical operations in laparoscopic environments are the natural fields of applications for this device.

Keywords: platform; spatial motion; remote actuation; cable actuation; laparoscopy; minimally-invasive surgery

1. Introduction

A parallel architecture offers several advantages to the designers of spatial platform mechanisms. In fact, they rely on a multi-loop topology that is certainly convenient for the stage stiffness and accuracy, whereas their forward kinematic analysis becomes more complex.

The Stewart–Gough platform, a mechanism that can be considered to be a milestone of parallel manipulators, was presented in 1965 [1]. This six-DoF (degree of freedom) system can be controlled in any combination by six motors and has the advantage of no fixed axis relative to the ground. Classical issues for parallel manipulators have been extensively studied in the literature, such as structural kinematics [2], workspace [3], singularity [4], optimal design [5], structure synthesis [6], manipulability [7], control of redundantly-actuated [8], and remotely cable-driven parallel manipulators [9]. On the other hand, more recent subjects have still not been well investigated for parallel platforms, such as connectivity and redundancy [10], topology [11], and planarity [12].

Classical parallel platforms are composed of rigid links and ordinary kinematic pairs with a geometric closure configuration. However, they could be built as compliant mechanisms [13–15], which present a series of advantageous characteristics: they are not subject to backlash and do not need lubrication; they can be also built from a unique block of material; and finally, they have generally a neutral or balance configuration from which the deformed poses can be achieved. In fact, they can reach a given configuration thanks to external forces that can be applied with a high precision to deform the elastic structure, using also a redundant driving strategy. More recently, active compliance [16,17] has been added as a further possibility in design, acting both as a series or in a parallel configuration with passive compliance.

These features make compliant mechanisms very interesting for applications where lubrication is impossible or where extreme accuracy is needed. For these reasons, a group multi-DoF compliant stages has also been developed in literature [18].

Since 1990, stiffness and conditioning maps of the workspace of parallel manipulators have been established [19], in order to prevent special types of singularities, which result in a loss of controllability. A six-DoF force sensor was designed in 1991 [20] on the base of the Stewart–Gough platform. In this layout, the fixed and mobile platforms are coupled by six spring-loaded pistons, whereas the length variations are measured by means of six linear voltage differential transformers mounted along the pistons. The Stewart–Gough inverse and forward kinematic transformations are used to calculate the forces and torques that are applied to the mobile platform. A three DoF translational compliant parallel platform was presented in 2005 [21] for nanomanipulation. Workspace, dexterity, and isotropic configurations were studied by using the pseudo rigid-body model (PRBM). A six-PSS (prismatic-spherical-spherical) nanopositioner compliant mechanism was designed [22] to be actuated by means of six multilayered piezoelectric actuators. The system is composed of one fixed plate, three two-PSS compliant mechanisms, and one end effector. The PRBM is also used in this investigation to study its kinematic analysis. A three DoF spatial translational accurate positioning compliant platform with flexible hinges and with piezoelectric actuators has been designed [23] to achieve high stiffness, a high speed of dynamic response, high kinematic accuracy, and high resolution. A six-DoF compliant parallel micro-scale manipulator with piezo-driven actuators and integrated force sensor has been designed [24] to provide real-time force information for feedback control. Kinematic and static analysis were investigated to achieve high positioning accuracy, compactness, and smooth and continuous displacements. A method for the optimal design and performance characterization of micromanipulators has been also applied to six-DOF parallel micromanipulators [25]. A three-DoF compliant platform has been studied [26] to make it able to move in three-dimensional space. The system was composed of compliant joints, actuators, and a central moving platform. The actuators consisted of three binary links, while the moving platform was an equilateral plate. The free end of each actuator and the central platform were connected by springs in such a way that the motion of the actuators was transmitted to the moving platform. Considering the applications at the micro scale and using the technology on which micro electro-mechanical systems (MEMS) are based, a compliant three-DoF plane parallel platform has been proposed [27] together with its kinetostatic optimization.

The "da Vinci" © surgical robot (by Intuitive Surgical, Inc., Sunnyvale, California, USA) is a widely-used system designed to facilitate surgical operations with a minimally-invasive approach. This robot makes use of multi-DoF end terminals as a resource to cope with different tasks, working in cooperation with robotic wrists or steerable instruments, and their characteristics have been extensively improved and refined throughout the years. For example, a 2 mm-diameter instrument equipped with a three-DoF wrist has been investigated and tested [28] for the robotic "da Vinci" platform. Moreover, new systems for measuring the end effector gripping force have also been developed by means of a torque transfer system [29]. Finally, deflection and force feedback from the "da Vinci" end effector have been provided by new types of strain gauges [30].

However, in most of the above-mentioned systems for remote manipulations that are widely adopted for minimally-invasive surgery, the wrists and end effector consist of mechanisms that have ordinary kinematic pairs, which are subject to backlash problems, and a serial kinematic structure, which is usually less robust than a parallel structure. The present paper presents a new six-DoF platform for remote manipulation that consists pf a mechanism with selective compliance (compliant mechanism) and has also a parallel structure. The device could be either part of "da Vinci" or independently driven by another positioning system. Originally, the system was conceived to work under a laparoscopic environment, and it is expected to improve its success in surgery, such as in laparoscopic sleeve gastrectomy (LSG) with cruroplasty [31], in surgical treatment of gastrointestinal stromal tumors of the duodenum [32], in colovesical fistula surgery with a minimally-invasive approach [33], in endoluminal loco-regional resection by transanal endoscopic microsurgery (TEM) [34–36], and also in low rectal anterior resection (LAR) [37].

2. Description of the New Concept Platform

In several applications, the position of a mobile platform in space has to be controlled by using remote means, such as cables. Usually, the conventional mechanisms that are able to implement such a feature are rather complicated because the platform has six degrees of freedom (DoF), and moreover, they are subject to backlash and parasitic deformations. This problem is quite general, but in this paper, a specific implementation suitable for surgical applications will be presented.

A new tendon-driven compliant mechanism is herein proposed, as a possible solution to the above-mentioned problem. The full mobility of the platform in space and, in particular, its raising motion is possible because of a new elastic joint, which combines the action of an elastic curved beam and of a portion of the conjugate surfaces. The invention consists basically of a wire-operated selective compliance mobile platform like the one represented in Figure 1. The mechanism base (k) is intended to be mounted on the end of a flexible tubular duct for endoscopic, surgical, therapeutic, or diagnostic uses.

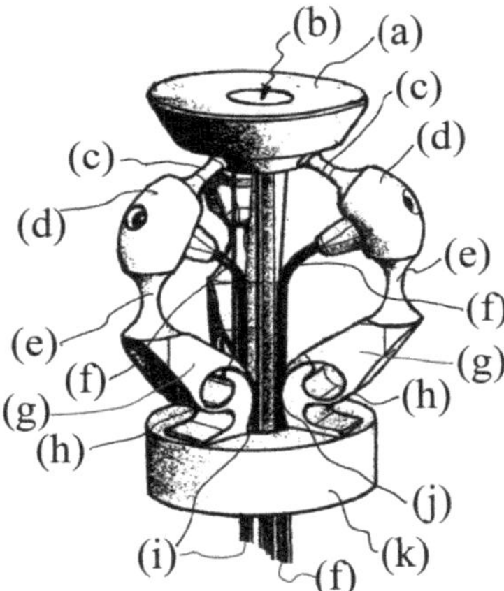

Figure 1. A view of the compliant platform and of all its elements: platform (**a**); plaform hole (**b**); type *S* elastic joints (**c**); type-U elastic joints (**e**); upper (**d**) and lower (**g**) links; upper (**f**) and lower cables (**i**); conjugate surface flexure hinge (CSFH) hinges (**h**); base link (**k**).

The platform (a) is designed for supporting surgical means, normally used during endoluminal operations, such as vision systems, forceps, scissors, cannulae, electrotomes, and so on. However, the system could be used for precision applications or in adverse environments, e.g., for operations in the aerospace environment. The base (k) and the platform (a) are connected through a plurality of legs, which provide mobility in space to the mobile surface (a), and are operated by actuating wires (i) and (f), running inside the tubular duct, to control and operate the platform remotely.

Since the mechanical structure of the mechanism is based on selective compliance and, particularly, on lumped compliance, there is a neutral configuration that is assumed when no external force is acting on the structure.

2.1. Deduction of the Pseudo-Rigid Body Equivalent Mechanism

A compliant mechanism can be generally modeled as its equivalent so-called pseudo-rigid body mechanism (PRBM). The PRBM is a mechanism with only ordinary kinematic pairs, which presents approximately the same motion as the original compliant mechanism in the neighborhood of the neutral configuration. For this purpose, each elastic joint is replaced by an ordinary kinematic pair, which can offer mobility that is compatible with the selective compliance characteristics of that elastic joint. Figure 2 shows the original compliant mechanism (Figure 2a) and its corresponding spatial PRBM (Figure 2b).

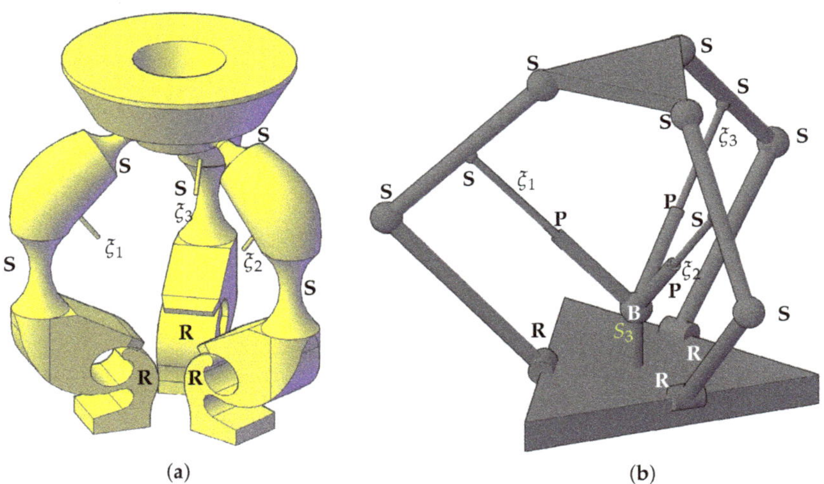

Figure 2. The compliant platform (**a**) and its corresponding pseudo-rigid body mechanism (**b**), with the equivalent revolute joints R, spherical joints S, prismatic pairs P, a triple spherical joint S_3, and the cable directions ξ_1, ξ_2, and ξ_3.

With reference to Figures 1 and 2, the elastic joints can be distinguished according to the following characteristics.

- A Type (e) elastic joint is delimited by a toroidal surface obtained by revolving a circular arc around a vertical axis γ_E. As a consequence, the neck cross-sectional area is normal to γ_E and is positioned in correspondence to the symmetry middle plane of the joint surface. The relative motion between the two connected pseudo-rigid parts has at least two degrees of freedom (DoF), namely those due to the bending along two orthogonal planes passing through γ_E. Furthermore, one more DoF could be added depending on whether the neck section allows the two opposite end sections to rotate significantly, one with respect to the other, about γ_E. For Type (e) joints, the choice of the diameter of the neck section is decisive for assuming whether the amount of torsional rotation might play a significant role. If so, Type (e) can be replaced by a three-DoF spherical S joint, otherwise a two DoF (Hooke's) universal U will be used. In the latter case, the flexure works approximately as a Cardan joint whose center is placed at the center of elastic weights of the elastic joint itself.

- A Type (c) elastic joint is also delimited by a portion of a toroidal surface obtained by the rotation of a circular arc around an axis of revolution $\gamma_C \neq \gamma_E$. For this type of joint, the rotations about γ_C are considered to be always non-negligible. In fact, they are provided with an even more reduced middle cross-section that gives rise to possible torsion rotations about γ_c. Therefore, the joints (c) substantially have the function of spherical elastic joints, the equivalent center of the spherical kinematic pair being placed at the center of elastic weights of the joint itself. For this reasons, Type (c) joints will be replaced by spherical joints S.

- Type (h) elastic joints belong to the class of the conjugate surface flexure hinge (CSFH). This kind of flexure has been patented (see Section 7) and extensively studied for several aspects: their theoretical behavior [38,39], design [40,41], dynamic simulation [42], fabrication [43–45], and applications [46–51]. Type (h) CSFH hinges selectively provide rotations around the CSFH axis with a very good accuracy, because of the presence of a portion of conjugate surfaces. Therefore, they are replaced by classical revolute joints R.

- Type (f) cables are the three moving elements for Type (d) upper links out of a total of six cables that are used to operate the six-DoF platform. For example, a simultaneous pull command on the three Type (f) cables will move the platform downward. The cables may be arranged in such a way so as to be aligned in the three directions ζ_1, ζ_2, and ζ_3, as illustrated in Figure 2a.

- Type (i) cables are the moving elements for Type (g) links. These three cables provide motion for the three CSFH hinges, and so, they regulate the orientation of Type (g) links. For example, a cable pull will induce a rotation that tends to raise the link upward.

Three linear actuators *SPS* are added to the system and positioned as in Figure 2b in order to replace the three actuated cables directed along the ζ_1, ζ_2, and ζ_3 directions. The three linear actuators are identified in the figure as the ones that are incident to the triple spherical joint S_3 positioned at Point B. This replacement is justified by the fact that the three cables are directed along lines that pass through the center B of the triple spherical joint S_3. This point is positioned in the middle of the fixed platform, in correspondence to a level that does not interfere with the CSFH hinges.

Finally, it is worth noting that the identification of a PRBM that adequately corresponds to the proposed compliant structure represents the first necessary step to further analyze the new systems, such as workspace and kinematic analysis, kinematic synthesis, and kinetostatic behavior.

2.2. Topological Analysis

Considering an equivalent mechanism PRBM with rigid bodies and classical U, S, and R pairs, three legs guarantee six degrees of freedom to the platform. In fact, the PRBM presents $\ell = 8$ rigid links and $m = 9$ kinematic pairs. In case Type (e) joints are replaced by U joints, the structure will be composed of three Type (h) revolute joints R with degree of constraint $c_i = 5$ in space, three Type (e) U joints with $c_i = 4$, and six S joints with $c_i = 3$. Therefore, according to the general topological Grubler's formula, the PRBM has:

$$F = \lambda\,(\ell - 1) - \sum_{i=1}^{m} c_i = 6\,(8 - 1) - 3 \cdot 5 - 3 \cdot 3 - 3 \cdot 4 = 6 \tag{1}$$

overall degrees of freedom (DoF), where $\lambda = 6$ is the mobility number for general spatial motion.

As mentioned above, torsion cannot always be excluded on U elastic Type (e) joints, depending on the minimum diameter of the normal cross-section, which would make them practically equivalent to S joints. In this case, Grubler's formula would yield $F = 9$ DoFs for the platform, but it must be remarked that three of such DoFs would be uniquely dedicated to providing idle rotations to the upper Type (d) links. In this case, any Link (d) would be connected to the remaining parts of the structure by means of two S type pairs, and so, rotations about an axis passing through the centers of the two spherical joints would be possible. These rotations are naturally counterbalanced by internal elastic reactions that would drive the system back toward the minimum of the potential elastic energy. Alternatively, they could make it easier to achieve an optimal attitude of Type (f) cables that pull down Type (d) links. For these reasons, a different manipulator could be obtained, as an alternative, where Type (e) links are replaced by Type (c) links, with no problems for mobility.

According to the result obtained by using Equation (1) and considering that each leg is supposed to be controlled by two cables, a three-legged platform will be the one that has a number of independent actuators that is equal to the number of DoF and therefore will be considered as an optimal choice for the compliant structure. In fact, according to the principle of selective compliance, any elastic element where compliance is concentrated presents a peculiar geometry that is intended for a specific deformation, for example a prescribed flexion or torsion axis, and all the deformations different from the prescribed ones are considered to be as parasitic deformations, because they do not contribute to the desired motion, but only increase the stress level in the elastic material. For this reason, a satisfying geometry of a compliant mechanism is the one that minimizes the parasitic deformations.

2.3. Direct Position Analysis

Thanks to the construction of the PRBM depicted in Figure 2b, it is possible to identify some fundamental characteristics of the platform, as they have been introduced in the literature. For example, it is clear that this platform is not a fully-parallel platform, because two DoF are needed for each leg. With reference to Figure 3a, the ith leg (with $i = 1, \ldots, 0$) will be composed of the pseudo rigid links $C_i A_i$ and $A_i P_i$, which respectively correspond to Type (g) and Type (d) links of the original compliant structure.

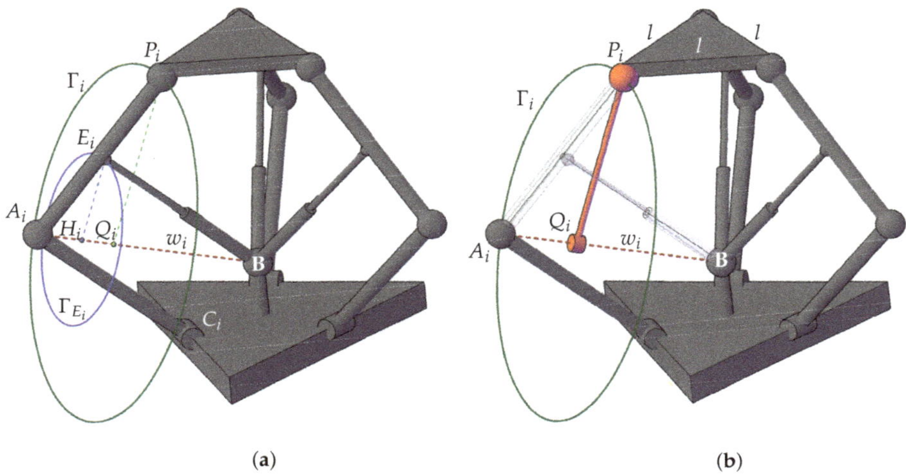

(a)   (b)

Figure 3. (**a**) The circles Γ_{E_i} and Γ_i as loci of points E_i and P_i, respectively, for the ith leg; (**b**) the virtual bar $Q_i P_i$ rotating along the axis w_i parallel to the line $A_i B$.

For the generic ith leg, one DoF is firstly required to define the angular position of Type (g) link $C_i A_i$. An angle α_i can be introduced to measure the rotation of $(C_i A_i)$ with respect to its position in the undeformed configuration of the CSFH joint. Since the latter corresponds to a revolute joint with a known rotation axis, the angle α_i uniquely identifies the position of the center A_i of the spherical pair. Secondly, another DoF is needed to assess the position of the upper Type (d) $A_i P_i$ link of the ith leg. Indeed, an assigned pull command on the Type (f) cable has the effect of bringing points E_i and B closer to each other. This effect could be similarly obtained by introducing in the PRBM a virtual linear actuator being active along the line BE_i. Any contraction of the linear actuator BE_i corresponds to a pull command of the (f) rope, while an extension of the linear actuator would represent a reduction of the cable tension. Since mobility is granted by the elasticity of the structure, its overall configuration will depend on the whole set of tensions that are assigned to the six cables. Furthermore, it is worth noticing that positive tensions (push) are not considered here, and therefore, the maximum elongation of the three virtual linear actuators will depend on the whole balance between the tensions that are applied to the cables and the structural elasticity. For the linear actuator BE_i, the pulled distance d_i can be introduced as the shortening of the original length $|BE_i|$ with respect to the distance between points B and E_i in the undeformed structure configuration.

Once the input parameters α_i and d_i have been assigned, the chain $BA_i E_i$ behaves as a rigid structure, which may rotate about the axis w_i passing through points A_i and B (see Figure 3b). Therefore, point E_i describes a circle Γ_{E_i} laying on a plane π_{H_i} that is orthogonal to w_i and whose center H_i is on line w_i. Since A_iP_i can be regarded as a rigid link, point P_i is also forced to lay on a circle Γ_i that lays on a plane π_{Q_i} parallel to π_{H_i}. The center Q_i of Γ_i is on the line w_i.

Considering the three loci Γ_i (with $i = 1, \ldots, 3$) of the possible positions of the points P_i, it is clear that three points of the upper platform, coincident with P_1, P_2, and P_3, will belong to the three circles Γ_1, Γ_2, and Γ_3, respectively. However, exactly the same constraint could be imposed by introducing a virtual link Q_iP_i, represented in Figure 3b, that rotates about the w_i axis. Therefore, three new links Q_iP_i, with $i = 1, \ldots, 3$, can be used to get rid of the three whole chains made of links C_iA_i, A_iP_i, and the linear actuators BE_i. After this substitution, a new three-legged structure is obtained, as depicted in Figure 4.

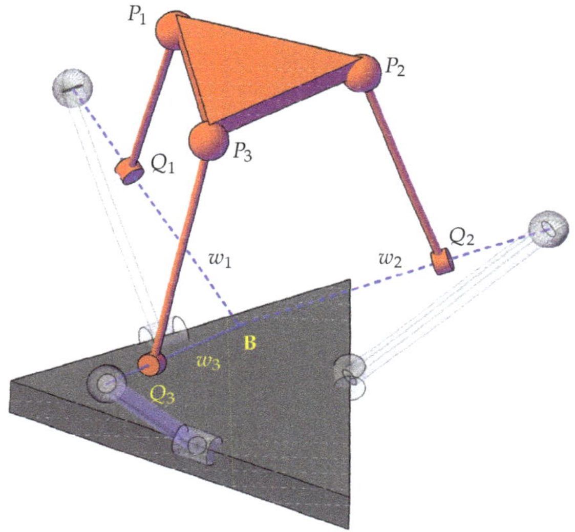

Figure 4. The reduced parallel structure: points Q_i, axes w_i and points P_i.

The new mechanism will have a null DoF because the whole set of input positions has already been assigned. In fact, after assigning a value to the three angles α_i and to the three displacements d_i, the axes w_i, the lengths $|Q_iP_i|$, and the positions of points Q_i can be uniquely identified. The resulting structure consists of a mobile platform that is connected to the base via three links only arranged in a parallel configuration, each one having one revolute R and one spherical S joint at its ends incident to the base and the platform, respectively. This structure has received the attention of several eminent scholars in the field and has been extensively studied [52–54]. From the geometric and kinematic point of view, the study of the possible assembly modes of this zero-DoF structure gives the same solution as for problem of finding the assembly modes of a six-DoF, so called 6-3-type fully-parallel mechanism, belonging to the class of Stewart platform mechanisms, once six elongation values are assigned to the six linear actuators along its legs.

Innocenti and Parenti Castelli solved the direct position analysis in 1990 [55] and found results that were coherent with Hunt's works [2]. According to this method (see also [56]), three closed-loop vector equations:

$$\overrightarrow{P_1P_2} = -\overrightarrow{Q_1P_1} - \overrightarrow{P_2Q_2} - \overrightarrow{Q_2Q_1}$$
$$\overrightarrow{P_2P_3} = -\overrightarrow{Q_2P_2} - \overrightarrow{P_3Q_3} - \overrightarrow{Q_3Q_2} \tag{2}$$
$$\overrightarrow{P_3P_1} = -\overrightarrow{Q_3P_3} - \overrightarrow{P_1Q_1} - \overrightarrow{Q_3Q_1}$$

can be written, where $\overrightarrow{P_1P_2}$, $\overrightarrow{P_2P_3}$, and $\overrightarrow{P_3P_1}$ are the edges of the upper platform.

For the sake of the present investigation $\overrightarrow{P_1P_2}$, $\overrightarrow{P_2P_3}$, and $\overrightarrow{P_3P_1}$ have constant modules because of the symmetry of the upper platform. Therefore, given the constant length l of the upper platform edges, Equation (2) can be rewritten as:

$$\overrightarrow{P_1P_2} \cdot \overrightarrow{P_1P_2} = l^2, \qquad \overrightarrow{P_2P_3} \cdot \overrightarrow{P_2P_3} = l^2, \qquad \overrightarrow{P_3P_1} \cdot \overrightarrow{P_3P_1} = l^2 \qquad (3)$$

from which three scalar conditions are obtained where the three unknown are the rotation angles θ_i (with $i = 1, \ldots, 3$) of the virtual bars Q_iP_i around the axes w_i.

It is now essential to remind that once the three rotations α_i of the CSFHs and the three displacements d_i of the Type (f) cables have been assigned, the zero-DoF structure illustrated in Figure 4 is completely configured because the positions of the points A_i, H_i, and Q_i can be easily calculated. Therefore, the solutions of the problem expressed by the set of Equation (3) are also the solution of the assembly configuration for the PRBM mechanism depicted in Figure 3a.

The system (3) consists of three second-order algebraic equations in three unknowns θ_i (with $i = 1, \ldots, 3$), where any equation contains two unknown variables only. Therefore, the solutions can be achieved by firstly eliminating one variable $\hat{\theta}$ from two equations where $\hat{\theta}$ appears, so obtaining one equation in the other two variables, and then, by eliminating one of the other two variables, say $\tilde{\theta}$, from the remaining two equations. The result is a 16th-order polynomial equation in the remaining variable θ, with 16 real and complex solutions for θ. Since for every solution θ, unique values of $\hat{\theta}$ and $\tilde{\theta}$ exist, a unique location of the upper platform is identified for each solution θ. These solutions correspond to the possible assembly configurations of the structure represented in Figure 4 and, as a consequence, of the PRBM depicted in Figure 3a, provided that the six inputs α_i and d_i have been assigned. However, for the sake of the present investigation, the only interesting solution will be the one compatible with rotations and displacements starting from the initial undeformed configuration represented in Figure 2a. This shows that the numerical approaches could be used more conveniently than the purely analytical ones, because they could converge to the actual configuration by using the undeformed pose as the starting guess. In the next section, finite element analysis (FEA) is applied to the compliant platform illustrated in Figure 2a, by assigning three different sets of input displacements.

3. Numerical Simulation of the Platform Pose

The peculiar geometry illustrated in Figure 1 is the result of a preliminary study that led to the definition of the patented structure. Three equal legs were positioned in a parallel configuration, each leg being actuated by two cables, and so, the whole structure was a tendon-driven mechanism. The geometry of each leg was chosen in such a way that the lower and upper cables induced a raising and lowering motion, respectively. The attachment points were also selected in order not to induce cable jamming. Further optimization of the attachment points and of the angles of the pseudo-rigid links will be the object of future investigations. Considering the nomenclature introduced in Figure 1, any moving cable (i) acts on a CSFH turning pair, relevantly operates the bending of the CSFH Type (h) flexure, and actually drives the rotations of Type (g) links. Cables (f) will have the effect of pulling the platform down because they are attached to the upper links. More in general, a configuration of the platform (a) will be defined by the whole set of forces applied to all the wires, since it depends on the interaction between the applied forces and the elasticity of joints.

Finite element analysis is a convenient way to assess the deformation of the whole structure for assigned values of the cable displacements. More specifically, FEA provided evidence that the adoption of the proposed platform can be a promising approach to find an efficient way for remote handling in several contexts.

Finite Element Analysis

Finite element analyses were conducted to evaluate the static behavior of the parallel platform. Three cases were considered depending on three different sets of input parameters and using $E = 1.1 \cdot 10^9$ Pa as the material Young's modulus.

Three cases will be studied:

I   pull on Type (f) cables and null tension on Type (i) cables;
II   pull on Type (i) cables and null tension on Type (f) cables;
III asymmetric pull: pull on one Type (f) cable and pull on two Type (i) cables;

The fixed reference frame (x, y, z) and the unit vectors $(\vec{u}_x, \vec{u}_y, \vec{u}_z)$ attached to the mobile platform were positioned as in Figure 5a. This figure also shows the input displacement d_1. Rotations θ_x, θ_y, and θ_z were assumed to be about $\vec{u}_x$, $\vec{u}_y$, and $\vec{u}_z$, respectively.

In the first case, a displacement was imposed on each one of the Type (f) cables, equal to $d_1 = d_2 = d_3 = 0.5$ mm, and null tensions on Type (i) cables, giving rise to null rotation angles $\alpha_1 = \alpha_2 = \alpha_3 = 0$ of the three CSFH hinges. The deformed configuration is reported in Figure 5b, showing a platform displacement along the z-axis in the negative direction. No other significant displacements or rotations were registered.

The second case corresponded to $\alpha_1 = \alpha_2 = \alpha_3 = 20°$ with $d_1 = d_2 = d_3 = 0$ mm. The deformed configuration is reported in Figure 6a, showing a platform displacement along the z-axis in the positive direction. Analogous to the previous case, no other significant displacements or rotations were registered.

The third case corresponded to a general actuation scheme with $\alpha_1 = 0°, \alpha_2 = 5°, \alpha_3 = 20°$, $d_1 = 1$ mm, $d_2 = d_3 = 0$ mm. The deformed configuration is reported in Figure 6b, showing a platform displacement with components along the $x-$ and y-axes and rotations about the x- and z-axes.

The input parameters, the platform displacement, and the rotations are reported in Table 1 for the three analyzed cases.

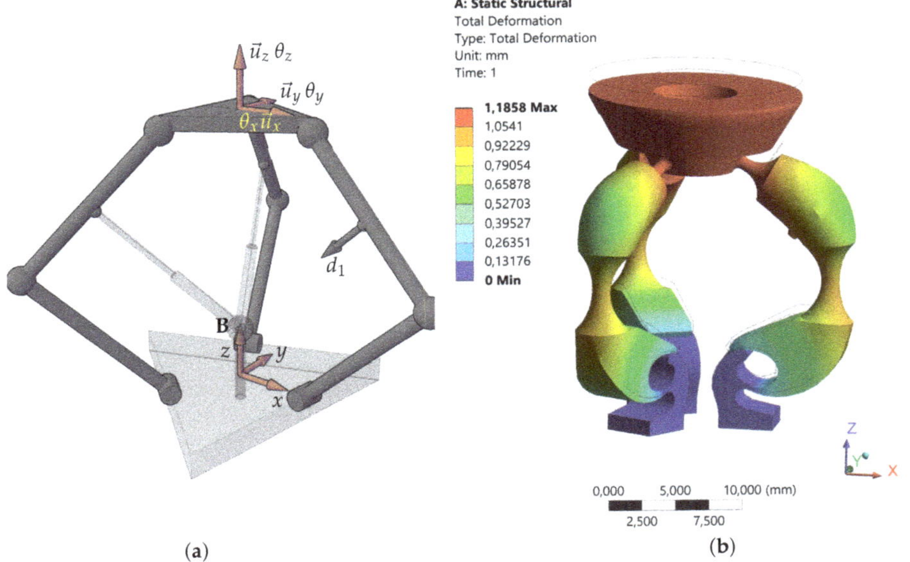

(a)   (b)

Figure 5. (a) Reference frames and nomenclature; (b) platform displacement along the z-axis, negative direction (Case 1).

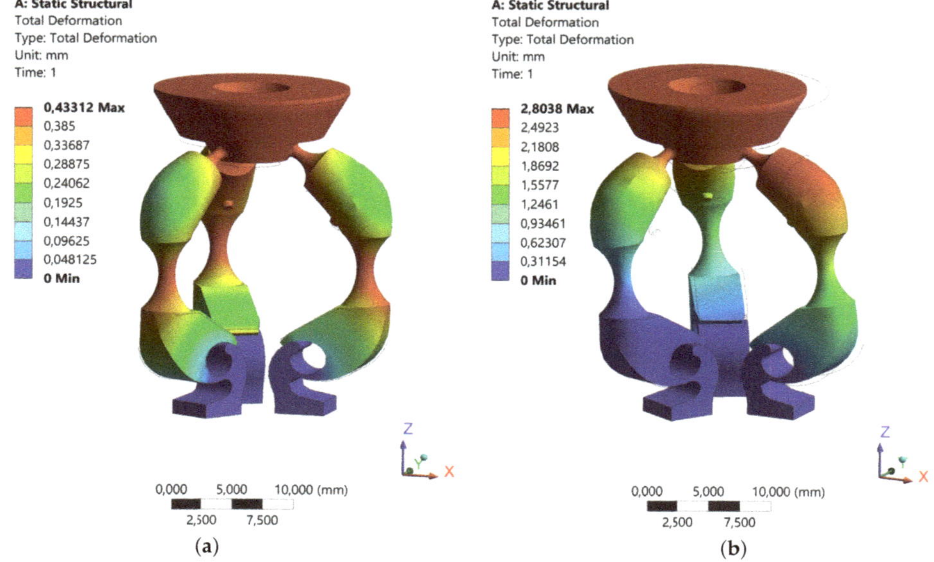

Figure 6. (**a**) Platform displacement along the z-axis, positive direction (Case 2); (**b**) platform pose for a generic actuation scheme (Case 3).

Table 1. Input rotations and displacements and their effect on the platform for the three analyzed cases.

Case 1		Case 2		Case 3	
Input	**Output**	**Input**	**Output**	**Input**	**Output**
$\alpha_1 = 0$	$u_x = 8 \cdot 10^{-3}$ mm	$\alpha_1 = 20°$	$u_x = -8 \cdot 10^{-3}$ mm	$\alpha_1 = 0°$	$u_x = -2.4$ mm
$\alpha_2 = 0$	$u_y = 2 \cdot 10^{-2}$ mm	$\alpha_2 = 20°$	$u_y = -1 \cdot 10^{-2}$ mm	$\alpha_2 = 5°$	$u_y = 0.1$ mm
$\alpha_3 = 0$	$u_z = -1.17$ mm	$\alpha_3 = 20°$	$u_z = 0.43$ mm	$\alpha_3 = 20°$	$u_z = -6 \cdot 10^{-2}$ mm
$d_1 = 0.5$ mm	$\theta_x = 8 \cdot 10^{-2}°$	$d_1 = 0$ mm	$\theta_x = 2 \cdot 10^{-2}°$	$d_1 = 1$ mm	$\theta_x = -1.7°$
$d_2 = 0.5$ mm	$\theta_y = 4 \cdot 10^{-2}°$	$d_2 = 0$ mm	$\theta_y = 2 \cdot 10^{-2}°$	$d_2 = 0$ mm	$\theta_y = -5 \cdot 10^{-2}°$
$d_3 = 0.5$ mm	$\theta_z = 9 \cdot 10^{-2}°$	$d_3 = 0$ mm	$\theta_z = 8 \cdot 10^{-4}°$	$d_3 = 0$ mm	$\theta_z = 1.1°$

4. Applications

Originally, the main activities for the invention herein presented were intended to be operated in the laparoscopic environment. Figure 7 shows a possible end-effector that could be mounted on the platform. With reference to the labels used in Figure 7, the endoluminal catheter (a) carries the base platform (b), which supports three Type (c) legs on which the mobile platform (d) is mounted. The mobile platform surface holds the surgical means (f) that can vary according to the type of required operation. Other miniaturized instruments can be mounted on the controlled platform, such as vision means (e) that allow the surgeon to observe the corporal cavity wherein the device is inserted. Moreover, a communication duct (g) was designed, allowing the surgeon to access the region to be operated. For this reason, the mobile and base platforms were also provided with an opening that connects the duct (g) with the tubular duct (a). With reference to Figure 7, it should be also noted that the surgical device was comprised of a shaped surface (f) that made it easier to act as a guiding element inside the cavity during the use of the device.

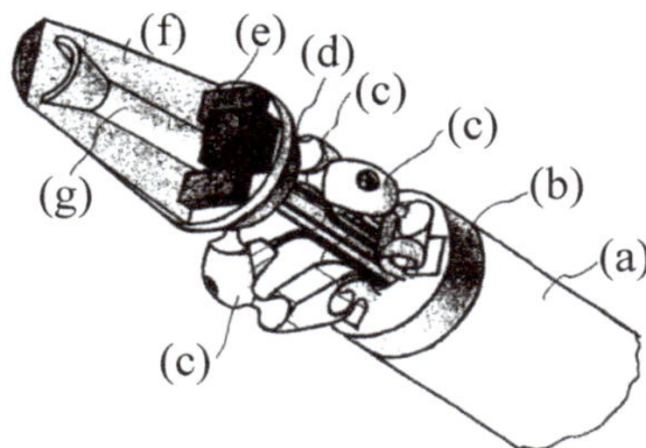

Figure 7. A view of an end-effector mounted in the platform: catheter (**a**); base link (**b**); legs (**c**); platform (**d**); on-board hardware (**e**); on-board end-effectors (**f**) and (**g**).

5. Discussion

The aim of this paper was to present a new concept high accuracy platform with spatial mobility, remotely actuated, and equipped with selective compliance joints. The system can be built on a unique material block or with a reduced number of components, which reduces the costs and time of production. It can be also built by means of 3D printing.

The platform was mounted on three arms, and it was remotely controlled by six wires. The system included a mobile plane and a connection base fixed to an endoscopic instrument. The instruments was accurate and maneuverable with continuous motion granted by the interaction between the cable tensions and the elastic energy stored in the elastic media. Based on selective compliance, the principle of design gave rise to optimal configurations where parasitic motions, friction, and wear were minimized. No lubrication was needed.

The invention is expected to drive surgical operations in laparoscopic, endogastric, or endarterial environments, but other applications are not excluded, such as in aerospace, automotive, appliances, and microelectronics. However, it is worth noting that the platform fabrication and its actual implementation for real operations deserve further refinements, and therefore, they will be treated in future contributions. Among the important subjects that still represent open problems, the following will be briefly addressed.

The first problem is the selection of an optimal material for the elastic parallel mechanism. This choice will depend on the overall size of the device and on the consequent preferred method of machining. In any case, some desired characteristics of a material that could be conveniently employed to build a compliant mechanism block were proposed in [57], among which are good performances at variable temperatures, biocompatibility, low susceptibility to creep, resistance to wear and fatigue, and the predictability of such properties. However, the most important characteristic is a high value of the ratio of the yield strength to Young's modulus. According to Table 2.1. presented in [57], there are several materials that have an acceptable ratio, among which are nylon, e-glass, Kevlar, and polyethylene.

The actual fabrication of the device will depend both on size and material. Unfortunately, the geometry is rather complicated, and so, many of the classical machining operations would be problematic.

The identification of a proper PRBM, as proposed in the present contribution, could be helpful to complete typical tasks in robotics, such as the study of the critical configurations. Furthermore, the PRBM will be useful to optimize the platform and the legs' geometry, in order to suit the required workspace, mobility, and mechanical advantage.

Finally, an efficient control algorithm would be required for each specific application.

6. Conclusions

A spatial, selective compliance, tendon-driven, and parallel platform was described and analyzed. This system was quite different from the compliant mechanisms previously presented either in the literature or the national patent institutions until today. Its mechanical structure was studied by identifying the equivalent pseudo-rigid body mechanism. The platform was also simulated with finite element analysis (FEA) by assigning different sets of cable tensions and obtaining results that matched well with the expected behavior. It is hoped that this article will offer new perspectives for general tasks of spatial body guidance and particularly in laparoscopic surgery.

7. Patents

The following patents describe the new concept platform and the related claims.

- N.P. Belfiore, M. Scaccia, F. Ianniello, M. Presta, L. Perfetti, Selective Compliance Wire Actuated Mobile Platform, particularly for Endoscopy Surgical Devices, Patent N0.2 US 8,845,520 B2, 30 September 2014.
- N.P. Belfiore, M. Scaccia, F. Ianniello, M. Presta, L. Perfetti, Selective Compliance Wire Actuated Mobile Platform, particularly for Endoscopy Surgical Devices, World Intellectual Property Organization, WO 2009/034552 A2, Int. Appl. No. PCT/IB2008/053698, Publ. Date 19 March 2009.
- N.P. Belfiore, M. Scaccia, F. Ianniello, M. Presta, Selective Compliance Hinge, US 8,191,204 B2, 5 June 2012.
- N.P. Belfiore, M. Scaccia, F. Ianniello, M. Presta, Selective Compliance Hinge, World Intellectual Property Organization, WO 2009/034551 A1, Int. Appl. No. PCT/IB2008/053697, Publ. Date 19 March 2009.

Funding: This research received no external funding.

Acknowledgments: The whole staff of the "Research and Technology Transfer Area" of the University of Rome la Sapienza is gratefully acknowledged. Particular thanks go to the Manager of the "Research and Technology Transfer Area" Antonella Cammisa, the Office Manager of the "Enhancement and Technology Transfer Office" Daniele Riccioni, and the Office Manager of the "Grant Office" Alessandra Intraversato, for their counseling and support during my long stay at Sapienza University. I am also grateful to several people from my new University of Roma Tre (my list would be too long to mention all of them) for supporting and encouraging me since my arrival in October 2017.

Conflicts of Interest: The author declares no conflict of interest.

Abbreviations

The following abbreviations are used in this manuscript:

CSFH	Conjugate surface flexure hinge
DoF	Degrees of freedom
MEMS	Micro electro-mechanical systems

References

1. Stewart, D. A Platform with Six Degrees of Freedom. *Proc. Inst. Mech. Eng.* **1965**, *180*, 371–386. [CrossRef]
2. Hunt, K.H. Structural Kinematics of in-parallel-actuated Robot-Arms. *J. Mech. Transm. Autom. Des.* **1983**, *105*, 705–712. [CrossRef]
3. Gosselin, C. Determination of the workspace of 6-DOF parallel manipulators. *J. Mech. Transm. Autom. Des.* **1990**, *112*, 331–336. [CrossRef]
4. Innocenti, C.; Parenti-Castelli, V. Singularity-free evolution from one configuration to another in serial and fully-parallel manipulators. *J. Mech. Des. Trans. ASME* **1998**, *120*, 73–79. [CrossRef]
5. Parenti-Castelli, V.; Di Gregorio, R.; Bubani, F. Workspace and optimal design of a pure translation parallel manipulator. *Meccanica* **2000**, *35*, 203–214. [CrossRef]

6. Fang, Y.; Tsai, L.W. Structure synthesis of a class of 4-DoF and 5-DoF parallel manipulators with identical limb structures. *Int. J. Robot. Res.* **2002**, *21*, 799–810. [CrossRef]
7. Merlet, J.P. Jacobian, manipulability, condition number, and accuracy of parallel robots. *J. Mech. Des. Trans. ASME* **2006**, *128*, 199–206. [CrossRef]
8. Cheng, H.; Yiu, Y.K.; Li, Z. Dynamics and Control of Redundantly Actuated Parallel Manipulators. *IEEE/ASME Trans. Mechatron.* **2003**, *8*, 483–491. [CrossRef]
9. Tang, X. An overview of the development for cable-driven parallel manipulator. *Adv. Mech. Eng.* **2014**, *2014*. [CrossRef]
10. Belfiore, N.P.; Di Benedetto, A. Connectivity and redundancy in spatial robots. *Int. J. Robot. Res.* **2000**, *19*, 1245–1261. [CrossRef]
11. Belfiore, N.P. Distributed Databases for the development of Mechanisms Topology. *Mech. Mach. Theory* **2000**, *35*, 1727–1744. [CrossRef]
12. Belfiore, N.P. Brief note on the concept of planarity for kinematic chains. *Mech. Mach. Theory* **2000**, *35*, 1745–1750. [CrossRef]
13. Howell, L.L.; Magleby, S.P.; Olsen, B.M. *Handbook of Compliant Mechanisms*; John Wiley and Sons: The Atrium Southern Gate, Terminus Road, Chichester, West Sussex, UK, 2013.
14. Howell, L.L.; Midha, A. Parametric deflection approximations for end-loaded, large-deflection beams in compliant mechanisms. *J. Mech. Des. Trans. ASME* **1995**, *117*, 156–165. [CrossRef]
15. Howell, L.L.; Midha, A. A method for the design of compliant mechanisms with small-length flexural pivots. *J. Mech. Des. Trans. ASME* **1994**, *116*, 280–290. [CrossRef]
16. Verotti, M.; Masarati, P.; Morandini, M.; Belfiore, N.P. Isotropic compliance in the Special Euclidean Group SE(3). *Mech. Mach. Theory* **2016**, *98*, 263–281. [CrossRef]
17. Verotti, M.; Belfiore, N.P. Isotropic compliance in E(3): Feasibility and workspace mapping. *J. Mech. Robot.* **2016**, *8*. [CrossRef]
18. Dunning, A.; Tolou, N.; Herder, J. Review Article: Inventory of platforms towards the design of a statically balanced six degrees of freedom compliant precision stage. *Mech. Sci.* **2011**, *2*, 157–168. [CrossRef]
19. Gosselin, C. Stiffness Mapping for Parallel Manipulators. *IEEE Trans. Robot. Autom.* **1990**, *6*, 377–382. [CrossRef]
20. Nguyen, C.C.; Antrazi, S.S.; Zhou, Z.L. Analysis and implementation of a 6 DOF Stewart platform-based force sensor for passive compliant robotic assembly. In Proceedings of the IEEE SOUTHEASTCON, Williamsburg, VA, USA, 7–10 April 1991; Volume 2, pp. 880–884.
21. Li, Y.; Xu, Q. Design and analysis of a new 3-DOF compliant parallel positioning platform for nanomanipulation. In Proceedings of the 5th IEEE Conference on Nanotechnology, Nagoya, Japan, 15 July 2005; Volume 2, pp. 133–136.
22. Wu, T.L.; Chen, J.H.; Chang, S.H. A six-dof prismatic-spherical-spherical parallel compliant nanopositioner. *IEEE Trans. Ultrason. Ferroelectr. Freq. Control* **2008**, *55*, 2544–2551.
23. Chen, Q.; Huang, Y.; Zhu, D.; Zhou, M. The design and finite element analysis of a compliant 3-DOF spatial translational ultra-precise positioning platform. In Proceedings of the IEEE International Conference on Intelligent Computing and Intelligent Systems, Xiamen, China, 29–31 October 2010; Volume 3, pp. 122–126.
24. Liang, Q.; Zhang, D.; Chi, Z.; Song, Q.; Ge, Y.; Ge, Y. Six-DOF micro-manipulator based on compliant parallel mechanism with integrated force sensor. *Robot. Comput.-Integr. Manuf.* **2011**, *27*, 124–134. [CrossRef]
25. Liu, X.J.; Wang, J.; Gao, F.; Wang, L.P. On the design of 6-DOF parallel micro-motion manipulators. In Proceedings of the IEEE International Conference on Intelligent Robots and Systems, Maui, HI, USA, 29 October–3 November 2001; Volume 1, pp. 343–348.
26. Correa, J.C.; Crane, C. Kinematic analysis of a three-degree of freedom compliant platform. *J. Mech. Des. Trans. ASME* **2013**, *135*. [CrossRef]
27. Belfiore, N.P.; Emamimeibodi, M.; Verotti, M.; Crescenzi, R.; Balucani, M.; Nenzi, P. Kinetostatic optimization of a MEMS-based compliant 3 DOF plane parallel platform. In Proceedings of the 9th International Conference on Computational Cybernetics, Tihany, Hungary, 8-10 July 2013; pp. 261–266.
28. Francis, P.; Eastwood, K.W.; Bodani, V.; Looi, T.; Drake, J. Design, Modelling and Teleoperation of a 2 mm Diameter Compliant Instrument for the da Vinci Platform. *Ann. Biomed. Eng.* **2018**, *46*, 1437–1449. [CrossRef] [PubMed]

29. Lee, C.; Park, Y.H.; Yoon, C.; Noh, S.; Lee, C.; Kim, Y.; Kim, H.C.; Kim, H.H.; Kim, S. A grip force model for the da Vinci end-effector to predict a compensation force. *Med Biol. Eng. Comput.* **2015**, *53*, 253–261. [CrossRef] [PubMed]

30. Pena, R.; Smith, M.J.; Ontiveros, N.P.; Hammond, F.L.; Wood, R.J. Printing Strain Gauges on Intuitive Surgical da Vinci Robot End Effectors. In Proceedings of the IEEE International Conference on Intelligent Robots and Systems, Madrid, Spain, 1–5 October 2018; Institute of Electrical and Electronics Engineers Inc.: Piscatawa, NJ, USA, 2018; pp. 806–812.

31. Balla, A.; Quaresima, S.; Ursi, P.; Seitaj, A.; Palmieri, L.; Badiali, D.; Paganini, A.M. Hiatoplasty with crura buttressing versus hiatoplasty alone during laparoscopic sleeve gastrectomy. *Gastroenterol. Res. Pract.* **2017**, *2017*. [CrossRef] [PubMed]

32. Popivanov, G.; Tabakov, M.; Mantese, G.; Cirocchi, R.; Piccinini, I.; D'Andrea, V.; Covarelli, P.; Boselli, C.; Barberini, F.; Tabola, R.; et al. Surgical treatment of gastrointestinal stromal tumors of the duodenum: A literature review. *Transl. Gastroenterol. Hepatol.* **2018**, *3*, 71. [CrossRef]

33. Cochetti, G.; Del Zingaro, M.; Boni, A.; Cocca, D.; Panciarola, M.; Tiezzi, A.; Gaudio, G.; Balzarini, F.; Ursi, P.; Mearini, E. Colovesical fistula: Review on conservative management, surgical techniques and minimally invasive approaches. *Giornale Chir.* **2018**, *39*, 195–207.

34. Quaresima, S.; Balla, A.; Dambrosio, G.; Bruzzone, P.; Ursi, P.; Lezoche, E.; Paganini, A.M. Endoluminal loco-regional resection by TEM after R1 endoscopic removal or recurrence of rectal tumors. *Minim. Invasive Ther. Allied Technol.* **2016**, *25*, 134–140. [CrossRef]

35. Paci, M.; Scoglio, D.; Ursi, P.; Barchetti, L.; Fabiani, B.; Ascoli, G.; Lezoche, G. Transanal Endocopic Microsurgery (TEM) in advanced rectal cancer disease treatment [Il ruolo della TEM nel trattamento dei tumori del retto extraperitoneale]. *Ann. Ital. Chir.* **2010**, *81*, 269–274.

36. Lezoche, E.; Fabiani, B.; D'Ambrosio, G.; Ursi, P.; Balla, A.; Lezoche, G.; Monteleone, F.; Paganini, A.M. Nucleotide-guided mesorectal excision combined with endoluminal locoregional resection by transanal endoscopic microsurgery in the treatment of rectal tumors: Technique and preliminary results. *Surg. Endosc.* **2013**, *27*, 4136–4141. [CrossRef]

37. Ursi, P.; Santoro, A.; Gemini, A.; Arezzo, A.; Pironi, D.; Renzi, C.; Cirocchi, R.; Di Matteo, F.; Maturo, A.; D'Andrea, V.; et al. Comparison of outcomes following intersphincteric resection vs low anterior resection for low rectal cancer: A systematic review. *G. Chir.* **2018**, *39*, 123–142.

38. Verotti, M. Effect of initial curvature in uniform flexures on position accuracy. *Mech. Mach. Theory* **2018**, *119*, 106–118. [CrossRef]

39. Verotti, M. Analysis of the center of rotation in primitive flexures: Uniform cantilever beams with constant curvature. *Mech. Mach. Theory* **2016**, *97*, 29–50. [CrossRef]

40. Sanò, P.; Verotti, M.; Bosetti, P.; Belfiore, N.P. Kinematic Synthesis of a D-Drive MEMS Device With Rigid-Body Replacement Method. *J. Mech. Des. Trans. ASME* **2018**, *140*. [CrossRef]

41. Verotti, M.; Dochshanov, A.; Belfiore, N.P. Compliance Synthesis of CSFH MEMS-Based Microgrippers. *J. Mech. Des. Trans. ASME* **2017**, *139*. [CrossRef]

42. Botta, F.; Verotti, M.; Bagolini, A.; Bellutti, P.; Belfiore, N.P. Mechanical response of four-bar linkage microgrippers with bidirectional electrostatic actuation. *Actuators* **2018**, *7*, 78. [CrossRef]

43. Veroli, A.; Buzzin, A.; Frezza, F.; de Cesare, G.; Hamidullah, M.; Giovine, E.; Verotti, M.; Belfiore, N.P. An approach to the extreme miniaturization of rotary comb drives. *Actuators* **2018**, *7*, 70. [CrossRef]

44. Bertini, S.; Verotti, M.; Bagolini, A.; Bellutti, P.; Ruta, G.; Belfiore, N.P. Scalloping and stress concentration in DRIE-manufactured comb-drives. *Actuators* **2018**, *7*, 57. [CrossRef]

45. Bagolini, A.; Ronchin, S.; Bellutti, P.; Chistè, M.; Verotti, M.; Belfiore, N.P. Fabrication of Novel MEMS Microgrippers by Deep Reactive Ion Etching with Metal Hard Mask. *J. Microelectromech. Syst.* **2017**, *26*, 926–934. [CrossRef]

46. Belfiore, N.P. Micromanipulation: A challenge for actuation. *Actuators* **2018**, *7*, 85. [CrossRef]

47. Potrich, C.; Lunelli, L.; Bagolini, A.; Bellutti, P.; Pederzolli, C.; Verotti, M.; Belfiore, N.P. Innovative silicon microgrippers for biomedical applications: Design, mechanical simulation and evaluation of protein fouling. *Actuators* **2018**, *7*, 12. [CrossRef]

48. Veroli, A.; Buzzin, A.; Crescenzi, R.; Frezza, F.; de Cesare, G.; D'Andrea, V.; Mura, F.; Verotti, M.; Dochshanov, A.; Belfiore, N.P. Development of a NEMS-technology based nano gripper. *Mech. Mach. Sci.* **2018**, *49*, 601–611.

49. Di Giamberardino, P.; Bagolini, A.; Bellutti, P.; Rudas, I.J.; Verotti, M.; Botta, F.; Belfiore, N.P. New MEMS tweezers for the viscoelastic characterization of soft materials at the microscale. *Micromachines* **2017**, *9*, 15. [CrossRef]

50. Cecchi, R.; Verotti, M.; Capata, R.; Dochshanov, A.; Broggiato, G.; Crescenzi, R.; Balucani, M.; Natali, S.; Razzano, G.; Lucchese, F.; et al. Development of micro-grippers for tissue and cell manipulation with direct morphological comparison. *Micromachines* **2015**, *6*, 1710–1728. [CrossRef]

51. Belfiore, N.P.; Simeone, P. Inverse kinetostatic analysis of compliant four-bar linkages. *Mech. Mach. Theory* **2013**, *69*, 350–372. [CrossRef]

52. Tsai, L.W. *Robot Analysis: The Mechanics of Serial and Parallel Manipulators*; John Wiley and Sons: New York, NY, USA, 1999.

53. Hunt, K.H. *Kinematic Geometry of Mechanisms*; Oxford University Press: Oxford, UK, 1979.

54. Merlet, J.P. *Parallel Robots*, 2nd ed.; Springer: Dordrecht, The Netherlands, 2006.

55. Innocenti, C.; Parenti-Castelli, V. Direct position analysis of the Stewart platform mechanism. *Mech. Mach. Theory* **1990**, *25*, 611–621. [CrossRef]

56. Parenti-Castelli, V.; Di Gregorio, R. Closed-form solution of the direct kinematics of the 6-3 type Stewart Platform using one extra sensor. *Meccanica* **1996**, *31*, 705–714. [CrossRef]

57. Howell, L.L. *Compliant Mechanisms*; John Wiley and Sons, Inc.: New York, NY, USA, 2001.

Article

Comparative Study of the Use of Different Sizes of an Ergonomic Instrument Handle for Laparoscopic Surgery

Juan A. Sánchez-Margallo [1,*], Alfonso González González [2], Lorenzo García Moruno [3], J. Carlos Gómez-Blanco [1], J. Blas Pagador [1] and Francisco M. Sánchez-Margallo [4]

[1] Bioengineering and Health Technologies Unit, Jesús Usón Minimally Invasive Surgery Centre, 10071 Cáceres, Spain; jcgomez@ccmijesususon.com (J.C.G.-B.); jbpagador@ccmijesususon.com (J.B.P.)
[2] Department of Mechanical, Energy, and Materials Engineering, University of Extremadura, 06800 Mérida, Spain; agg@unex.es
[3] Department of Graphical Expression, University of Extremadura, 06800 Mérida, Spain; lgmoruno@unex.es
[4] Scientific Direction, Jesús Usón Minimally Invasive Surgery Centre, 10071 Cáceres, Spain; msanchez@ccmijesususon.com
* Correspondence: jasanchez@ccmijesususon.com

Received: 31 December 2019; Accepted: 19 February 2020; Published: 24 February 2020

Abstract: Previous studies have shown that the handle design of laparoscopic instruments is crucial to surgical performance and surgeon's ergonomics. In this study, four different sizes of an ergonomic laparoscopic handle design were tested in a blind and randomized fashion with twelve surgeons. They performed three laparoscopic tasks in order to analyze the influence of handle size. Execution time, wrist posture, and finger and palm pressure were evaluated during the performance of each task. The results show a significant reduction in the time required to complete the eye-manual coordination task using the appropriate handle. The incorrectly sized handle resulted in a rise in palm pressure and a reduction in the force exerted by the thumb during the transfer task. In the hand-eye coordination task, the use of the right handle size led to an increase in middle finger pressure. In general, surgeons had an ergonomically adequate wrist flexion in all tasks and an acceptable radio-ulnar deviation during the transfer task using the ergonomic instrument handle. Surgeons found it comfortable the use of the ergonomic handle. Therefore, the use of an appropriately sized instrument handle allows surgeons to improve ergonomics and surgical performance during the laparoscopic practice.

Keywords: laparoscopy; instrument handle; ergonomics; hand pressure; wrist posture; parametric design

---

1. Introduction

Laparoscopic surgery has become a standard procedure instead of open approach for many surgical interventions and it is widely used in health services around the world. This rising interest in laparoscopy is based on several benefits, such as better surgical outcomes for patients and resource optimization for healthcare services [1]. Furthermore, this surgical technique has constantly evolved over the past few years, mainly due to training efforts of surgeons [2–4] and technological advances of surgical equipment [5]. However, there are some important technical limitations that should be addressed in order to increase both performance and wellbeing of surgeons: reduced freedom of movements due to the fixed location of the surgical ports, the need for high precision through long instruments, loss of tactile feedback (no direct touch of intracorporeal anatomical structures), and the use of bidimensional vision instead of real stereoscopic (3D) vision. The combination of these features of laparoscopy makes it difficult for surgeons to achieve hand-eye coordination, depth perception and performance of certain intracorporeal maneuvers, resulting in prolonged forced postures and the consequent impairment of surgeons' performance and precision and the potential onset of musculoskeletal disorders [6]. In order to overcome some of these problems, ergonomic criteria should be applied to both patients and surgeons [7,8]. In this respect, an important field of application of ergonomics for surgeons is the improvement of surgical equipment and tools, as well as its adaptation to the surgeon's needs. The design of ergonomically better surgical tools would reduce the surgeon's muscle fatigue and other associated diseases [9], with the potential improvement of the surgical performance [7]. This results in enhanced patient safety and surgical results.

In view of the above, an ergonomic laparoscopic instrument handle has been designed and developed in previous studies [10,11]. This new surgical tool follows the design criteria for universal objects named "Design for All" criteria [12]. These criteria were included in the Universal Design guide that enumerates the seven principles of a universal design [13]. One of these principles is to facilitate "flexibility in use" so that most people could use the designed products [14,15]. In this regard, hand tools must meet these design criteria in order to overcome obstacles such as age, gender, psychomotor skills or laterality.

An important challenge to be dealt with is to ensure a consistent tool design, both in shape and dimensions, for a specific functionality [16]. Stoklasek et al. demonstrated the onset of pain and discomfort after long use of hand tools with a poor ergonomic design that eventually led to fingers' numbness and, in some cases, paresthesia [17]. Additionally, anthropometric dimensions of hand tools is another fundamental design aspect. Hand size is a critical factor for precision tools, specifically in laparoscopic instruments [18]. Surgeons with small or large hands often have problems gripping laparoscopic tools, mainly because of the size or shape of the instrument handle. For this reason, they sometimes have to grip these instrument handles in a different way to that the designers intended [18]. It is important to note that laparoscopic surgical tools are commonly sold in a standard size [19–21].

Specifically, the opening and closing mechanisms of surgical instruments present complications for surgeons with large and small hands. Surgeons with small hands often have difficulties in using instruments with a bigger size than the ideal one, mainly for instruments with a power grip [16]. For this reason, several studies have focused on addressing this problem by analyzing the opening and closing system of laparoscopic instruments [22], force applied during its use [23], hand and fingers kinematics [24,25], and EMG activity during the laparoscopic practice [26–28]. Changes in the handle design of laparoscopic instruments have been shown to affect localized muscle fatigue, mainly in the muscle groups of the surgeons' forearms [6,29–31].

Therefore, the objective of this work is to carry out a blind study to analyze the effects of different sizes of the ergonomic design of a laparoscopic handle on the surgeon's ergonomics and surgical performance. This will be done by comparing the use of a handle that is the appropriate size for the surgeon's hand with one that is the wrong size. The study will be carried out during the performance of various basic training tasks for laparoscopic surgery.

2. Materials and Methods

2.1. Description of the New Hhandle Design

Figure 1 shows a functional prototype of the patented laparoscopic instrument handle (EP10382362) used to carry out this study. This intellectual property resulted from the ERGOLAP project (Obtaining Criteria for Ergonomic Design of Laparoscopic Surgery Instruments. DPI2007-65902-C03-03) developed by the University of Extremadura (UEX), Jesús Usón Minimally Invasive Surgery Centre (JUMISC) and Institute of Biomechanics of Valencia (IBV). This novel instrument handle design is mainly characterized by a power grip that keeps the precision of movements. The parameters of the laparoscopic instrument handle design in relation to the anthropometry of the surgeon's hand are described in [10,11]. This design provides the surgeon with a neutral posture of the wrist thanks to the wide contact surface between the palm and the handle, promoting the reduction in fatigue and possible associated musculoskeletal disorders.

Figure 1. Design of the laparoscopic instrument handle (patent EP2471473A1) developed in the ERGOLAP project.

Additionally, the parametric 3D CAD design allowed us to easily scale the model to fit different hand sizes in an ergonomic way using the Palm Length Measured (PLM) [11]. Therefore, four different sizes (XS, S, M and L) (Figure 2) were 3D printed for this study following the specifications defined in previous studies [10,11].

Figure 2. Prototypes of the handles for laparoscopic instruments developed. The sizes of the handles, from left to right, are XS, S, M and L.

2.2. Development of the Prototypes

As described above, the four instrument handles were 3D printed with PLA using a Prusa Original MK3 MMUS2.5 (Prusa Research s.r.o., Prague, Czech Republic), together with the triggers to control the opening and closing of the instrument tip and the wheels to rotate it. In order to obtain fully functional prototypes, these components were mounted on the internal mechanism of a standard laparoscopic forceps. This allows surgeons to test these new handle designs during the execution of conventional laparoscopic tasks.

2.3. Participants

Twelve surgeons participated in this study. Two surgeons were left-handed. All participants performed the tasks using the new design of the instrument handle with their dominant hand and a conventional laparoscopic instrument with their non-dominant hand. Four participants were novice surgeons (<10 laparoscopic procedures performed), 5 had an intermediate level of experience in laparoscopic surgery (between 10 and 100 laparoscopic procedures) and 3 were experienced surgeons (>100 laparoscopic procedures).

2.4. Tasks

At the beginning of the study, the size category (XS, S, M and L) of each participant's hand was identified using a template. Each task was performed using both an instrument handle with the correct size for the surgeon and a handle with a difference of two sizes (wrong size) in a blind and randomized fashion. The type of laparoscopic instrument used with both the left and right hands to perform the activities was grasping forceps. For this study, participants were asked to perform three different basic laparoscopic tasks on a physical simulator (SIMULAP®; CCMIJU, Cáceres, Spain) with a 10-mm, 30-degree rigid laparoscope (Karl Storz GmbH & Co. KG, Tuttlingen, Germany) as the vision system (Figure 3). The height of the table and monitor were adapted to the comfort needs of each participant and the laparoscopic camera was fixed to prevent movements during the execution of the tasks. Surgeons were given a brief explanation of how to operate the controls on the new instrument handle (trigger and wheel). Since the study was randomized, participants received no training period with the instrument in order to prevent them from being familiar with a specific handle size and thus avoid possible bias.

Figure 3. Tasks: (**a**) Hand-eye coordination, (**b**) coordinated traction and (**c**) transfer task (labyrinth).

2.4.1. Hand-Eye Coordination

To perform this task, participants were asked to grasp three colored objects from one side of the pegboard with their closest hand, transfer the object mid-air to their opposite hand, and place the object on a peg on the other side of the pegboard (Figure 3a). Once all three objects had been transferred, the process was reversed.

2.4.2. Coordinated Traction

This task required the placement of an elastic band from a peg placed on one side of the pegboard to three other pegs on the other side, sequentially (Figure 3b). The starting peg was on the opposite side of the participant's dominant hand.

2.4.3. Transfer Task (Labyrinth)

Participants were asked to transfer a straight needle through a circuit of rings, distributed at different angles (Figure 3c). The needle was driven through the rings using the dominant hand and with the support of the non-dominant hand. The order of the rings and orientation of the needle driving were indicated with numbers and arrows on the pegboard, respectively.

2.5. Assessment

The execution time was recorded for each trial. In addition, ergonomic aspects of using the new instrument handle design during the execution of each task, such as wrist posture and pressure exerted on the instrument handle, were analyzed (Figure 4).

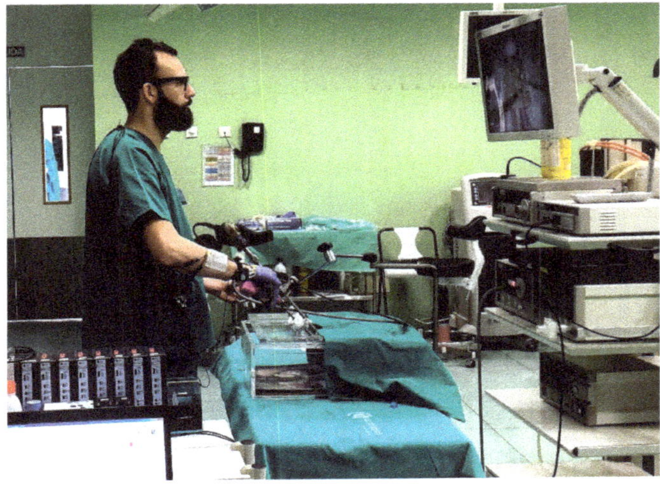

Figure 4. Setup of the study. A surgeon is performing the transfer task using an instrument handle with a size L.

2.5.1. Hand Pressure

To measure the fingers and hand pressure while using the instrument handle, the FingerTPS^{TM} (Pressure Profile Systems, Inc., Hawthorne, CA, USA) wearable and wireless system was employed. In this study, a set of five sensors to record the pressure exerted by the surgeon's distal phalanges of the thumb and index, middle and ring fingers, as well as the palm, were used. The system was calibrated for each participant.

2.5.2. Wrist Posture

The flexion-extension and radio-ulnar deviation of the wrist were recorded using a dual-axis electrogoniometer (Biopac Systems, Inc., Goleta, CA, USA) attached to the surgeon's hand and forearm using medical adhesive tape. In order to measure both axes simultaneously, two DA100C amplifiers (Biopac Systems, Inc.) were used. This device was calibrated at the beginning of the study. The risk level of the wrist posture was determined by means of a modified version of the RULA (Rapid Upper Limb Assessment) method as described in [32]. Risk values were defined according to the wrist angle, with wrist flexion-extension and radio-ulnar deviation angles between −15° and 15° being considered ergonomically acceptable.

2.5.3. Subjective Assessment

At the end of each task, the participants were asked to complete a questionnaire to evaluate, using a visual scale from 1 to 10, the comfort of performing the task with each instrument handle used.

2.6. Statistical Analysis

For statistical analysis, the Wilcoxon signed rank test was used to compare the surgeons' performance and ergonomic data of both study groups (right and wrong instrument handle size).

All statistical analyses were carried out using R version 3.6.1 (R Foundation for Statistical Computing, Vienna, Austria). The results are shown as mean and standard deviation. For all tests, $p < 0.05$ was considered statistically significant.

3. Results

3.1. Execution Time

The surgeons required significantly less time to complete the hand-eye coordination task using the adequate size of instrument handle (Figure 5). The execution time was similar for the other tasks, regardless of the type of instrument handle used.

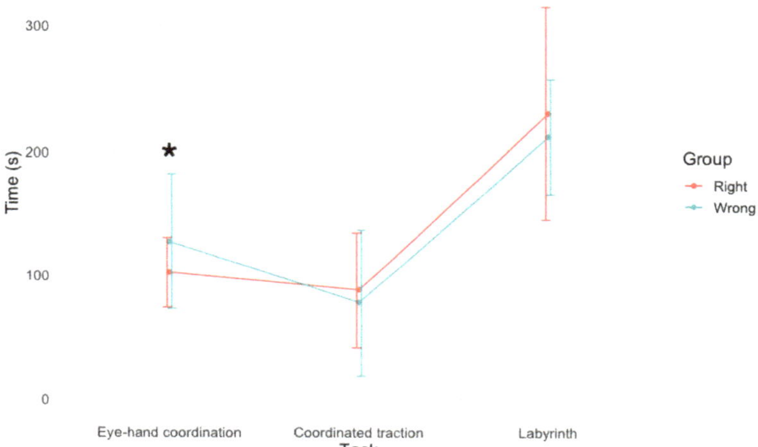

Figure 5. Execution time of the three laparoscopic tasks using the right and wrong instrument handle size. * $p < 0.05$.

3.2. Hand Pressure

There was a remarkable increase in palmar pressure on the wrong size instrument handle for the coordinated traction and transfer tasks compared to the right size handle, being statistically significant for the latter task (Figure 6). Using the right size of instrument handle led surgeons to apply more pressure with the thumb and middle fingers during the transfer and hand-eye coordination tasks, respectively, when compared to the use of the wrong size handle.

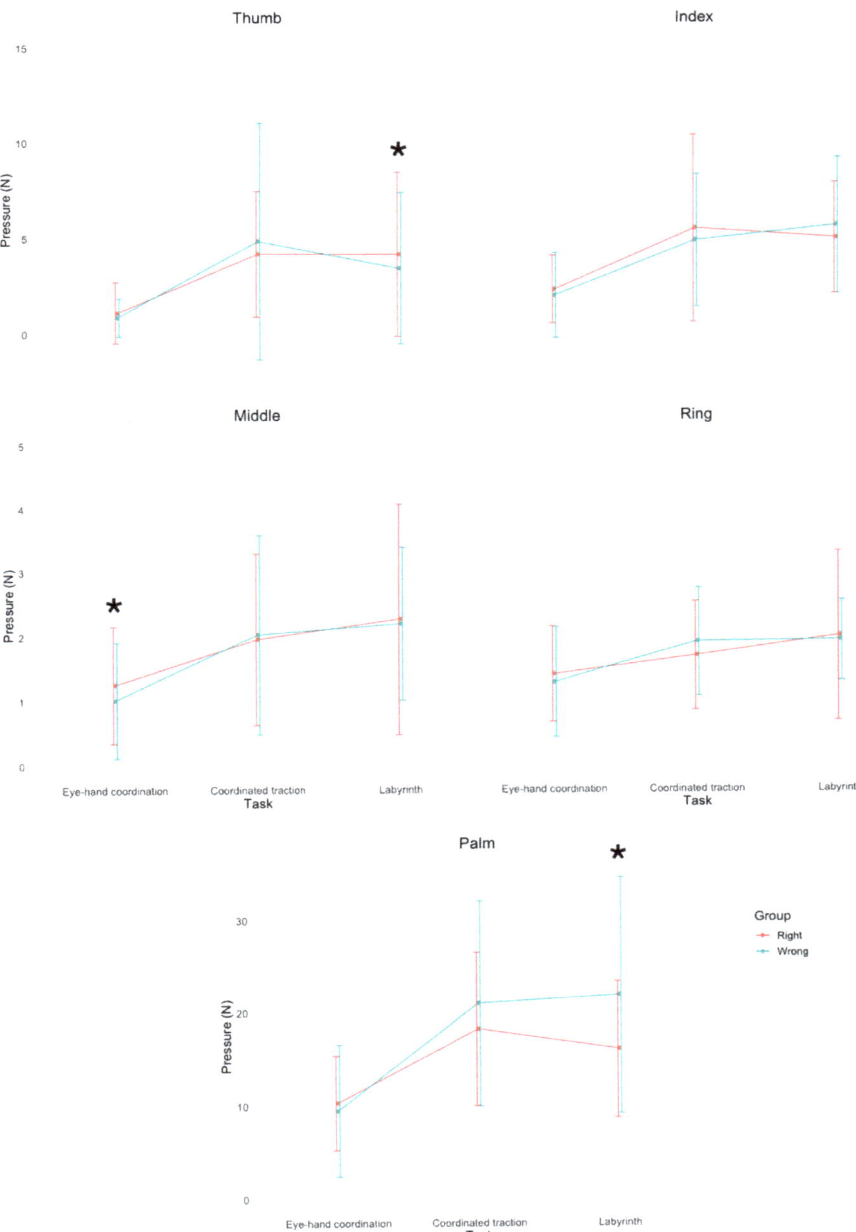

Figure 6. Pressure exerted by the surgeons' fingers (thumb, index, middle and ring fingers) and palm during the laparoscopic tasks using the right and wrong instrument handle size. The range of pressure values (y axis) has been adapted to each data set. * $p < 0.05$.

3.3. Wrist Posture

No statistically significant differences in wrist posture with each instrument handle were shown for any of the laparoscopic tasks (Figures 7 and 8). Regarding the wrist radio-ulnar deviation, the surgeons only obtained an ergonomically acceptable wrist posture, according to the criteria of RULA method, during the transfer task and for both sizes of instrument handle. In the case of wrist flexion-extension, both groups of handles led to an adequate posture of the wrist for all the tasks.

3.4. Subjective Assessment

The results from the questionnaires show that there were no statistically significant differences in the comfort of using the two different sizes of instrument handle for the three performed tasks (Figure 9).

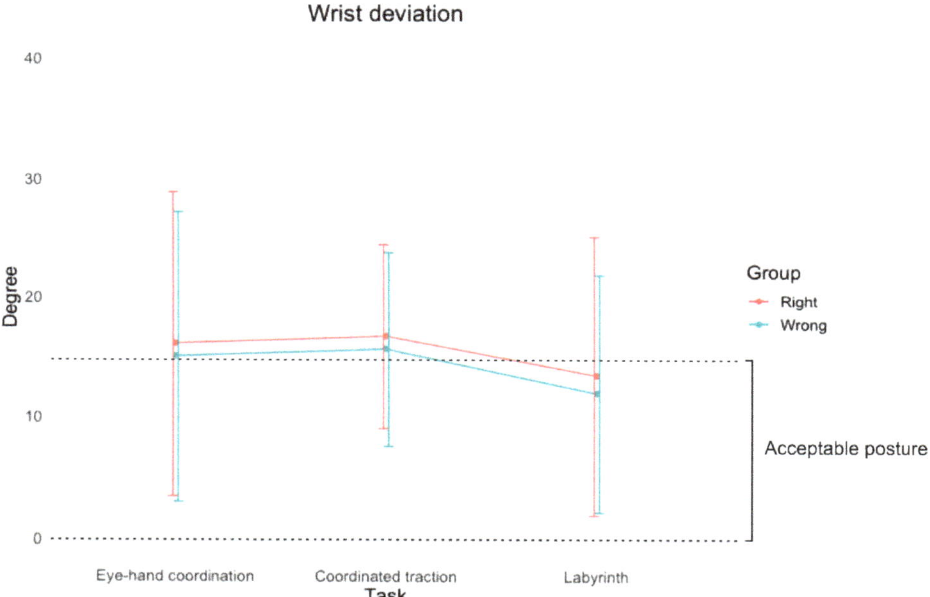

Figure 7. Radio-ulnar deviation of the surgeons' wrist during the laparoscopic tasks using the right and wrong size of instrument handle. The graph indicates the acceptable posture of the wrist according to RULA method.

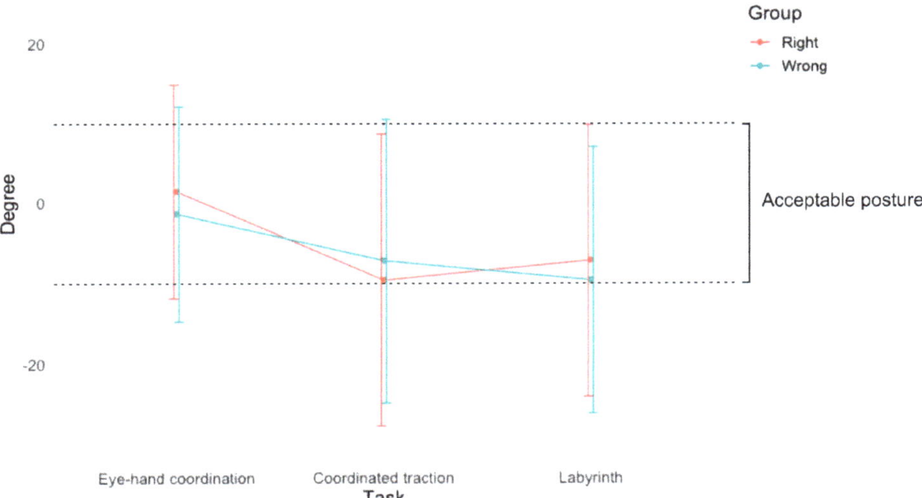

Figure 8. Flexion-extension of the surgeons' wrist during the laparoscopic tasks using the right and wrong size of instrument handle. The graph indicates the acceptable posture of the wrist according to RULA method.

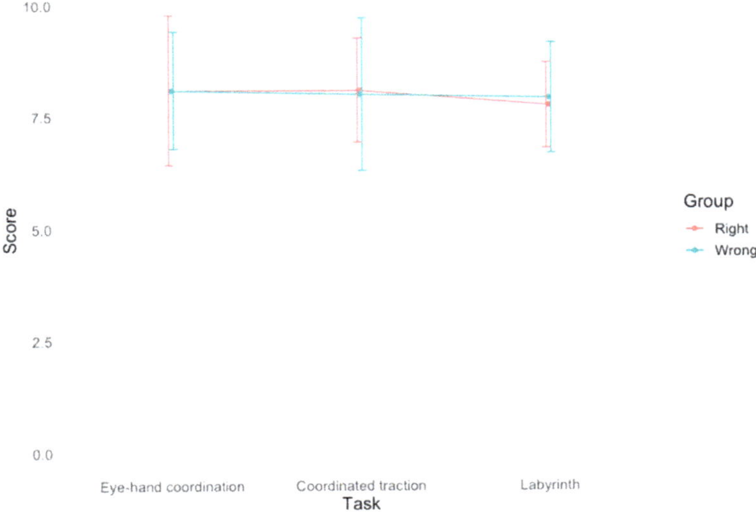

Figure 9. Results of the surgeons' evaluation regarding the comfort of performing each task using the two types of instrument handles.

4. Discussion

There are several studies in the scientific literature that confirm the need to improve the comfort and ergonomic criteria of instruments for laparoscopic surgery. The instrument handle is an essential part of laparoscopic instruments, whose inappropriate design can have a detrimental effect on the surgeon's efficiency and wellbeing [33]. The rate of work-related musculoskeletal disorders in laparoscopy are from 73% to 100%, with poor instrument handle design being one of the main risk factors [34,35]. In a survey by Santos-Carreras et al., they assessed the acceptance of different surgical instrument handle designs and they reported that conventional laparoscopic instrument handles such as the axial handle scored lowest in terms of comfort [36]. Other studies also indicated that axial handles require significantly more muscle activity than the other common types of instrument handle [37]. In this study, we have analyzed the relationship between the surgeons' ergonomics and performance and different sizes of a novel laparoscopic instrument handle design [11]. The results verify that the use of different instrument handle sizes has consequences on the surgeon's grip pressure and surgical skills.

It seems that the use of an adequate size of instrument handle helps hand-eye coordination and bimanual dexterity in the development of laparoscopic tasks, reducing the execution time, as shown by the results of this study. Tung et al. also presented an innovative laparoscopic handle design with a pistol grip attempting to address some of the ergonomic limitations in laparoscopic surgery [38]. Although the validation of this handle design was mainly subjective, they reported significantly shorter execution times in performing cutting and peg transfer tasks when compared to a conventional laparoscopic tool. However, execution time may not be a very meaningful assessment parameter in the case of the coordinated traction task. This task requires participants to exert a high degree of traction with the instruments and, therefore, to firmly the instrument handle. Due to the metal material of the rings, they sometimes slipped off the instrument tip, thus increasing the time required to complete the task, despite the surgeon's skills.

Regarding the pressure exerted by the surgeon's fingers and palm using the instrument handle, the palm exerted the highest amount of pressure on the handle when compared to the rest of the anatomical sites analyzed. This result was also reported by Rossi et al. using a power grip configuration [39]. In the case of the fingers, the thumb and index finger applied a greater amount of pressure than the middle and ring fingers. This may be because of the primary use of the thumb to control the wheel to rotate the instrument tip and the index finger to operate the trigger that opens and closes it. It should be noted that the use of an inappropriate handle size, especially in more complex tasks such as the transfer task (labyrinth), led to a significant increase in the pressure on the palm. It seems that this substantial increase in palmar pressure was accompanied by a reduction in the pressure exerted by the thumb when controlling the instrument wheel. Therefore, instead of distributing the force exerted by the fingers and the palm in a more balanced way, in this case it was mainly focused on the surgeon's palmar area. In a power grip configuration, like the one we have in our design, the handle should be held by the partially bent fingers and the palm, exerting a counter pressure with the thumb lying near the plane of the palm. Excessive strain and repetitive hand actions associated with power gripping contribute to trauma and localized fatigue [39].

The instrument handle design presented in this study allowed for an ergonomically acceptable wrist flexion-extension posture during the laparoscopic practice. This posture was also maintained for all the different handle sizes evaluated in the study. Other studies, using commercially available instruments [40] and prototypes of instrument handles [41,42], also reported that a pistol handle configuration provides an ergonomically adequate wrist flexion, providing a significantly better wrist posture than using a conventional laparoscopic tool. The results of the presented study are consistent with those presented in a previous study in which we evaluated another ergonomic power grip handle for a handheld robotic laparoscopic instrument [43]. The ergonomic analysis showed that, after training, the use of the robotic instrument resulted in an ergonomically acceptable flexion–extension of the wrist, in contrast to the posture acquired by using the conventional needle holder with axial handle. In addition, in another study we analyzed the surgeon's wrist flexion-extension posture when handling conventional instruments for laparoscopic surgery, including axial, curved, and ringed handles [44]. The results showed that, for all instruments, most surgeons have ergonomically unfavorable angles of wrist flexion, which were prone to possible musculoskeletal disorders. One of the first ergonomic studies of laparoscopic instrument handles by Matern et al. also warned of ergonomic deficiencies in wrist flexion when using ring handles and excessive radial deviation in axial handles [45]. In the case of the wrist radio-ulnar deviation, we observed the most notable improvements using the ergonomic instrument handle during the execution of the transfer task. These results reinforce the ergonomic benefits to the surgeon of using the presented handle design for laparoscopic practice.

The general design of the handle has been positively evaluated by the participants of this study with regard to the comfort of use, without showing significant differences between the groups of handle sizes. Perhaps in this study we have been conservative and a difference of only two sizes is not enough to appreciate noticeable changes in the use of the presented instrument handle design. For future studies, we will seek to compare designs with larger size differences, which could be also a very common situation in surgical practice, mainly between female and male surgeons.

This study presents some limitations of which we are aware of. Although there are other studies in the same field with a similar or smaller number of subjects [37,40], the number of participants in this study is restricted. In further studies, we will increase the study population in order to obtain more representative results and allow comparison by experience levels in laparoscopic surgery. The skills acquired by experienced surgeons in handling laparoscopic instruments may be an influential factor in their ergonomics when using different sizes of instrument handles. On the other hand, in this study we only analyzed the instrument handles with laparoscopic grasping forceps. Future work will be carried out on the use of this handle design for other laparoscopic instruments, such as a needle holder, and thus analyze the experience of surgeons in more complex tasks as intracorporeal suturing.

5. Conclusions

The results of this study show that the ergonomic design of the instrument handle improves the surgeon's ergonomics and comfort during laparoscopic practice, promoting a more neutral operating wrist posture. The use of an adequate size of the instrument handle allows for improved surgical performance, surgeons required less time to complete the eye-hand coordination task than using the incorrect size, and ergonomics by reducing the contact pressure on the surgeon's palm during the transfer task.

Author Contributions: Conceptualization, J.A.S.-M., A.G.G., L.G.M., J.B.P., J.C.G.-B., F.M.S.-M.; methodology, J.A.S.-M., A.G.G., L.G.M., J.B.P., J.C.G.-B., F.M.S.-M.; software, A.G.G., J.C.G.-B., J.A.S.-M.; validation, J.A.S.-M.; formal analysis, J.A.S.-M.; investigation, J.A.S.-M., A.G.G., J.C.G.-B.; resources, F.M.S.-M. and L.G.M.; data curation, J.A.S.-M., J.B.P., A.G.G.; writing—original draft preparation, J.A.S.-M., J.B.P., A.G.G.; writing—review and editing, F.M.S.-M. and L.G.M.; visualization, J.A.S.-M.; supervision, F.M.S.-M. and L.G.M.; project administration, F.M.S.-M. and L.G.M.; funding acquisition, F.M.S.-M., A.G.G. and L.G.M. All authors have read and agreed to the published version of the manuscript.

Funding: This work has been partially funded by Junta de Extremadura (Spain) and FEDER funds (GR18199, GR18029, GR18176).

Acknowledgments: The authors wish to acknowledge the support of this research work to the VI Regional Research Plan and to the Regional Government of Extremadura and the European Regional Development Fund (ERDF) linked to the financial of the research groups Manufacturing Engineering (GR18029), Jesús Usón Minimally Invasive Surgery Centre (GR18199) and INNOVA (GR18176). The authors appreciate the excellent willingness to cooperate of the surgeons who participated in the study.

Conflicts of Interest: The authors declare no conflict of interest.

References

1. Tozzi, F.; Berardi, G.; Vierstraete, M.; Kasai, M.; de Carvalho, L.A.; Vivarelli, M.; Troisi, R.I. Laparoscopic versus open approach for Formal right and left hepatectomy: A propensity score matching analysis. *World J. Surg.* **2018**, *42*, 2627–2634. [CrossRef]
2. Sánchez-Margallo, F.M.; Pérez-Duarte, F.J.; Azevedo, A.M.; Sánchez-Margallo, J.A.; Sánchez-Hurtado, M.A.; Díaz-Güemes, I. An Analysis of Skills Acquisition During a Training Program for Experienced Laparoscopists in Laparoendoscopic Single-Site Surgery. *Surg. Innov.* **2014**, *21*, 320–326.

3. Sánchez-Peralta, L.F.; Sánchez-Margallo, F.M.; Moyano-Cuevas, J.L.; Pagador, J.B.; Enciso, S.; Gómez-Aguilera, E.J.; Usón-Gargallo, J. Learning curves of basic laparoscopic psychomotor skills in SINERGIA VR simulator. *Int. J. Comput. Assist. Radiol. Surg.* **2012**, *7*, 881–889.

4. Sánchez-Margallo, J.A.; Sánchez-Margallo, F.M.; Pagador, J.B.; Gómez, E.J.; Sánchez-González, P.; Usón, J.; Moreno, J. Video-based assistance system for training in minimally invasive surgery. *Minim. Invasive Ther. Allied Technol.* **2011**, *20*, 197–205. [CrossRef]

5. Büchel, D.; Mårvik, R.; Hallabrin, B.; Matern, U. Ergonomics of disposable handles for minimally invasive surgery. *Surg. Endosc.* **2010**, *24*, 992–1004. [CrossRef]

6. Berguer, R.; Gerber, S.; Kilpatrick, G.; Remler, M.; Beckley, D. A comparison of forearm and thumb muscle electromyographic responses to the use of laparoscopic instruments with either a finger grasp or a palm grasp. *Ergonomics* **1999**, *42*, 1634–1645. [CrossRef]

7. Park, A.; Lee, G.; Seagull, F.J.; Meenaghan, N.; Dexter, D. Patients benefit while surgeons suffer: an impending epidemic. *J. Am. Coll. Surg.* **2010**, *210*, 306–313. [CrossRef]

8. Manasnayakorn, S.; Cuschieri, A.; Hanna, G. Ideal manipulation angle and instrument length in hand-assisted laparoscopic surgery. *Surg. Endosc.* **2008**, *22*, 924–929. [CrossRef]

9. Sari, V.; Nieboer, T.E.; Vierhout, M.E.; Stegeman, D.F.; Kluivers, K.B. The operation room as a hostile environment for surgeons: physical complaints during and after laparoscopy. *Minim Invasive Ther. Allied Technol.* **2010**, *19*, 105–109. [CrossRef]

10. González, A.G.; Salgado, D.R.; Moruno, L.G. Optimisation of a laparoscopic tool handle dimension based on ergonomic analysis. *Int. J. Ind. Ergonom.* **2015**, *48*, 16–24. [CrossRef]

11. González, A.G.; Salgado, D.R.; García Moruno, L.; Sánchez Ríos, A. An ergonomic customized-tool handle design for precision tools using additive manufacturing: a case study. *Appl. Sci.* **2018**, *8*, 1200. [CrossRef]

12. Mace, R. Universal design: Barrier free environments for everyone. *Des. West* **1985**, *33*, 147–152.

13. Mace, R.L.; Ostroff, E.; Connell, B.R.; Jones, M.; Mueller, J.; Mullick, A.; Sanford, J.; Steinfeld, E.; Follette Story, M.; Vanderheiden, G. *The Principles of Universal Design*; NC State University: Raleigh, ND, USA, 1997.

14. Story, M.F.; Mueller, J.L.; Mace, R.L. *The Universal Design File: Designing for People of all Ages and Abilities*; NC State University: Raleigh, ND, USA, 1998.

15. Clarkson, P.J.; Coleman, R. History of Inclusive Design in the UK. *Appl. Ergon.* **2015**, *46*, 235–247. [CrossRef] [PubMed]

16. Berguer, R.; Hreljac, A. The relationship between hand size and difficulty using surgical instruments: a survey of 726 laparoscopic surgeons. *Surg. Endosc.* **2004**, *18*, 508–512. [CrossRef] [PubMed]

17. Stoklasek, P.; Mizera, A.; Manas, M.; Manas, D. Improvement of handle grip using reverse engineering, CAE and Rapid Prototyping. In Proceedings of the MATEC Web of Conferences 20th International Conference on Circuits, Systems, Communications and Computers (CSCC 2016), Corfu Island, Greece, 14–17 July 2016.

18. Sutton, E.; Irvin, M.; Zeigler, C.; Lee, G.; Park, A. The ergonomics of women in surgery. *Surg. Endosc.* **2014**, *28*, 1051–1055. [CrossRef] [PubMed]

19. Kong, Y.K.; Lowe, B.; Lee, S.J.; Krieg, E. Evaluation of handle design characteristics in a maximum screwdriving torque task. *Ergonomics* **2007**, *50*, 1404–1418. [CrossRef] [PubMed]

20. Harih, G.; Dolšak, B. Tool-handle design based on a digital human hand model. *Int. J. Ind. Ergonom.* **2013**, *43*, 288–295. [CrossRef]

21. Harih, G.; Dolšak, B. Comparison of subjective comfort ratings between anatomically shaped and cylindrical handles. *Appl. Ergon.* **2014**, *45*, 943–954. [CrossRef]

22. Maithel, S.; Villegas, L.; Stylopoulos, N.; Dawson, S.; Jones, D. Simulated laparoscopy using a head-mounted display vs traditional video monitor: an assessment of performance and muscle fatigue. *Surg. Endosc.* **2005**, *19*, 406–411. [CrossRef]

23. Horeman, T.; Dankelman, J.; Jansen, F.W.; van den Dobbelsteen, J.J. Assessment of laparoscopic skills based on force and motion parameters. *IEEE Trans. Biomed. Eng.* **2013**, *61*, 805–813. [CrossRef]

24. Pérez-Duarte, F.J.; Lucas-Hernández, M.; Matos-Azevedo, A.M.; Sánchez-Margallo, J.A.; Díaz-Güemes, I.; Sánchez-Margallo, F.M. Objective analysis of surgeons' ergonomy during laparoendoscopic single-site surgery through the use of surface electromyography and a motion capture data glove. *Surg. Endosc.* **2014**, *28*, 1314–1320. [CrossRef]

25. Loukas, C.; Georgiou, E. Multivariate autoregressive modeling of hand kinematics for laparoscopic skills assessment of surgical trainees. *IEEE Trans. Biomed. Eng.* **2011**, *58*, 3289–3297. [CrossRef]

26. Pérez-Duarte, F.J.; Sánchez-Margallo, F.M.; Martín-Portugués, I.; Sánchez-Hurtado, M.A.; Lucas-Hernández, M.; Sánchez-Margallo, J.A.; Usón-Gargallo, J. Ergonomic Analysis of Muscle Activity in the Forearm and Back Muscles During Laparoscopic Surgery. *Surg. Laparosc. Endosc. Percutan. Tech.* **2013**, *23*, 203–207. [CrossRef]

27. Szeto, G.P.; Poon, J.T.; Law, W.L. A comparison of surgeon's postural muscle activity during robotic-assisted and laparoscopic rectal surgery. *J. Robot. Surg.* **2013**, *7*, 305–308. [CrossRef]

28. Dufaug, A.; Barthod, C.; Goujon, L.; Forestier, N. Ergonomic surgical practice analysed through sEMG monitoring of muscular activity. In Proceedings of the 2016 IEEE 18th International Conference on e-Health Networking, Applications and Services (Healthcom), Munich, Germany, 14–17 September 2016; pp. 1–6.

29. Quick, N.; Gillette, J.; Shapiro, R.; Adrales, G.; Gerlach, D.; Park, A. The effect of using laparoscopic instruments on muscle activation patterns during minimally invasive surgical training procedures. *Surg. Endosc.* **2003**, *17*, 462–465. [CrossRef]

30. Emam, T.; Frank, T.; Hanna, G.; Cuschieri, A. Influence of handle design on the surgeon's upper limb movements, muscle recruitment, and fatigue during endoscopic suturing. *Surg. Endosc.* **2001**, *15*, 667–672. [CrossRef]

31. Emam, T.; Hanna, G.; Cuschieri, A. Ergonomic principles of task alignment, visual display, and direction of execution of laparoscopic bowel suturing. *Surg. Endosc.* **2002**, *16*, 267–271. [CrossRef]

32. Sánchez-Margallo, F.M.; Sánchez-Margallo, J.A. Assessment of Postural Ergonomics and Surgical Performance in Laparoendoscopic Single-Site Surgery Using a Handheld Robotic Device. *Surg. Innov.* **2018**, *25*, 208–217. [CrossRef]

33. Li, Z.; Wang, G.; Tan, J.; Sun, X.; Lin, H.; Zhu, S. Building a framework for ergonomic research on laparoscopic instrument handles. *Int. J. Surg.* **2016**, *30*, 74–82. [CrossRef]

34. Catanzarite, T.; Tan-Kim, J.; Whitcomb, E.L.; Menefee, S. Ergonomics in Surgery: A. Review. *Female Pelvic. Med. Reconstr. Surg.* **2018**, *24*, 1–12. [CrossRef]

35. Lucas-Hernández, M.; Pagador, J.B.; Pérez-Duarte, F.J.; Castelló, P.; Sánchez-Margallo, F.M. Ergonomics problems due to the use and design of dissector and needle holder: A survey in minimally invasive surgery. *Surg. Laparosc. Endosc. Percutan. Tech.* **2014**, *24*, e170–e177. [CrossRef] [PubMed]

36. Santos-Carreras, L.; Hagen, M.; Gassert, R.; Bleuler, H. Survey on Surgical Instrument Handle Design. *Surg. Innov.* **2012**, *19*, 50–59. [CrossRef] [PubMed]

37. Matern, U.; Kuttler, G.; Giebmeyer, C.; Waller, P.; Faist, M. Ergonomic aspects of five different types of laparoscopic instrument handles under dynamic conditions with respect to specific laparoscopic tasks: An electromyographic-based study. *Surg. Endosc.* **2004**, *18*, 1231–1241. [CrossRef] [PubMed]

38. Tung, K.D.; Shorti, R.M.; Downey, E.C.; Bloswick, D.S.; Merryweather, A.S. The effect of ergonomic laparoscopic tool handle design on performance and efficiency. *Surg. Endosc.* **2015**, *29*, 2500–2505. [CrossRef] [PubMed]

39. Rossi, J.; Berton, E.; Grélot, L.; Barla, C.; Vigouroux, L. Characterisation of forces exerted by the entire hand during the power grip: effect of the handle diameter. *Ergonomics* **2012**, *55*, 682–692. [CrossRef]

40. Yu, D.; Lowndes, B.; Morrow, M.; Kaufman, K.; Bingener, J.; Hallbeck, S. Impact of novel shift handle laparoscopic tool on wrist ergonomics and task performance. *Surg. Endosc.* **2016**, *30*, 3480–3490. [CrossRef]

41. Sancibrian, R.; Gutierrez-Diez, M.C.; Torre-Ferrero, C.; Benito-Gonzalez, M.A.; Redondo-Figuero, C.; Manuel-Palazuelos, J.C. Design and evaluation of a new ergonomic handle for instruments in minimally invasive surgery. *J. Surg. Res.* **2014**, *188*, 88–99. [CrossRef]

42. Van Veelen, M.A.; Meijer, D.W.; Uijttewaal, I.; Goossens, R.H.M.; Snijders, C.J.; Kazemiere, G. Improvement of the laparoscopic needle holder based on new ergonomic guidelines. *Surg. Endosc.* **2003**, *17*, 699–703. [CrossRef]

43. Sánchez-Margallo, J.A.; Sánchez-Margallo, F.M. Initial experience using a robotic-driven laparoscopic needle holder with ergonomic handle: assessment of surgeons' task performance and ergonomics. *Int. J. Comput. Assist. Radiol. Surg.* **2017**, *12*, 2069–2077. [CrossRef]

44. Sánchez-Margallo, F.M.; Sánchez-Margallo, J.A.; Pagador, J.B.; Moyano, J.L.; Moreno, J.; Usón, J. Ergonomic Assessment of Hand Movements in Laparoscopic Surgery Using the CyberGlove. In *Computational Biomechanics for Medicine*; Miller, K., Nielsen, P.M.F., Eds.; Springer: New York, NY, USA, 2010; pp. 121–128.
45. Matern, U.; Waller, P. Instruments for minimally invasive surgery: Principles of ergonomic handles. *Surg. Endosc.* **1999**, *13*, 174–182. [CrossRef]

Article

High-Frequency Deep Sclerotomy, A Minimal Invasive Ab Interno Glaucoma Procedure Combined with Cataract Surgery: Physical Properties and Clinical Outcome

Bojan Pajic [1,2,3,4,*], **Zeljka Cvejic** [2], **Kaweh Mansouri** [5,6], **Mirko Resan** [4] **and Reto Allemann** [1]

[1]   Eye Clinic Orasis, Swiss Eye Research Foundation, 5734 Reinach AG, Switzerland; rallemann@gmail.com
[2]   Department of Physics, Faculty of sciences, University of Novi Sad, Trg Dositeja Obradovica 4,
      21000 Novi Sad, Serbia; zeljka.cvejic@df.uns.ac.rs
[3]   Division of Ophthalmology, Department of Clinical Neurosciences, Geneva University Hospitals,
      1205 Geneva, Switzerland
[4]   Faculty of Medicine of the Military Medical Academy, University of Defense, 11000 Belgrade, Serbia;
      resan.mirko@gmail.com
[5]   Glaucoma Center, Montchoisi Clinic, 1000 Lausanne, Switzerland; kwmansouri@gmail.com
[6]   Department of Ophthalmology, University of Colorado, Denver, CO 80014, USA
[*]   Correspondence: bpajic@datacomm.ch; Tel.: +41-62-765-60-80

Received: 25 November 2019; Accepted: 23 December 2019; Published: 27 December 2019

Abstract: Background: The efficiency and safety of primary open-angle glaucoma with high-frequency deep sclerotomy (HFDS) combined with cataract surgery has to be investigated. Methods: Right after cataract surgery, HFDS was performed ab interno in 205 consecutive patients with open angle glaucoma. HFDS was performed with a custom-made high-frequency disSection 19 G probe (abee tip 0.3 × 1 mm, Oertli Switzerland). The bipolar current with a frequency of 500 kHz is applied. The nasal sclera was penetrated repetitively six times through the trabecular meshwork and consecutively through Schlemm's canal. Every time, a pocket of 0.3 mm high and 0.6 mm width was created. Results: Mean preoperative intraocular pressure (IOP) was 24.5 ± 2.1 mmHg (range 21 to 48 mmHg). After 48 months, the follow up average IOP was 15.0 ± 1.7 mmHg (range 10 to 20 mmHg). Postoperative IOP has been significantly reduced compared to preoperative IOP for all studied cases ($p < 0.001$). After 48 months, the target IOP less than 21 mmHg reached in 84.9%. No serious complications were observed during the surgical procedure itself and in the postoperative period. Conclusions: HFDS is a minimally invasive procedure. It is a safe and efficacious surgical technique for lowering IOP combined with cataract surgery.

Keywords: glaucoma; ab interno; minimally invasive glaucoma surgery; cataract surgery; high-frequency deep sclerotomy; intraocular pressure

---

1. Introduction

It is well known that glaucoma comes first in cases of irreversible vision loss and is the second leading cause of blindness worldwide [1]. Nowadays in developed countries, less than 50% of people are unaware of their diagnosis, mainly because of the asymptomatic nature of chronic glaucoma [2].

Trabeculectomy remains the gold standard for glaucoma surgery despite high rates of complications [3–5]. Choroidal effusions, hypotony, shallow anterior chambers, and hyphema are known as early postoperative complications [6–8]. Late complications are often bleb-related, including leakage, blebitis, and endophthalmitis. These complications are more common if antimetabolites like 5-Fluorouracil and Mitomycin C are used [9].

Outflow resistance is caused primarily by the juxtacanilicular trabecular meshwork and, especially, by the inner wall of the Schlemm's canal. This fact is used for the concept of the trabecular meshwork bypass [10]. It is presumed that 35% of the outflow resistance arises in the inner wall of the Schlemm's canal [11].

In addition to the procedures already mentioned, we focus in this work on an enhanced Schlemms canal (SC), which can be addressed both internally and externally. According to the current classification, the ab interno methods include trabectome, iStent, Hydrus, the results presented in this work, and high-frequency deep sclerotomy (HFDS) [12]. The trabectome was introduced in 2004 and has an electroablation of an arc of trabecular meshwork where direct access to the collector channels is given. The intraocular pressure (IOP) reduction of up to 40% is stated in the literature for this ab interno method [13,14]. Another approach for a micro-invasive glaucoma surgery (MIGS) from interno is the iStent. With this device, the aqueous humor is drained directly into the Schlemm's canal, thus avoiding the resistance of the trabecular meshwork [15]. Another representative of the ab interno procedure of the Schemm's canal microstent is the so-called hydrus. It is an 8 mm long crescent-shaped open structure that the bend and sludge canal adapts. It is introduced into the Schlemm's canal by a clear corneal incision [16]. The HFDS is an internally operative glaucoma method that was introduced in 1999. Initially, this surgical procedure was known as sclerothalamotomy ab interno (STT ab interno), and its name was only changed to HFDS in 2012. The aim of the new nomenclature was to describe the procedure more precisely. Regarding the procedure, there were no differences between STT ab interno and HFDS [17–20].

The aim of all MIGS is that, in addition to being effective, they can be carried out simply, quickly and inexpensively with the longest possible effect. The aim of this study is to show the clinical 4-year investigations of this trans-trabecular surgery using the Schlemm's canal as an outflow pathway combined with cataract surgery.

2. Materials and Methods

In this single center study from December 2012 till December 2016, 223 patients with insuffiently controlled primary open-angle glaucoma under maximal tolerated topical therapy without history of prior ocular surgery with significant cataract were included. Exlcusion criteria were monophthalmia, angle closure with or without glaucoma, missing willingness to attend follow-up examinations with randomization or any psychiatric disorder, and any condition that affects the optical system (severe alteration of cornea, anterior chamber, or retina). During the observation period of 48 months, 18 patients did not finish the follow up due to several reasons, so finally 205 patients were included in the study. HFDS was performed combined with cataract surgery, with cataract surgery first being performed. Both procedures were always performed by the same surgeon. A complete ophthalmological examination was carried out in each patient prior to surgery. Further ophthalmologic follow-up examinations were carried out postoperatively on days 1, 2, and 3, after 1 and 4 weeks, and then every 3 months until 48 months. Gonioscopy with a three-mirror goniolens was performed after 4 weeks to check the persistence of the sclerotomy. Tobramycin/dexamethasone and pilocarpin 2% eye drops were applied 3 times daily for 1 month. The tenets of the Declaration for Helsinki were followed in this study. The cantonal ethics committee of Aargau approved the study. According to the ethics committee, patient consent is not required for this retrospective study.

High-Frequency Diathermic Probe

The high-frequency diathermic probe (abee® Glaucoma Tip, Oertli Instrumente AG, Berneck, Switzerland) consists of an inner and outer coaxial electrode. Both electrodes are isolated, while the inner one is made by platinum.

The dimensions of platinum probe tips are 1 mm in length, 0.3 mm high, and 0.6 mm wide. The platinum probe tip is bent posteriorly at an angle of 150°. The probe external diameter is 0.9 mm. Electrical current is modulated at MF (~500 kHz) frequency. The temperature of approximately 130 °C at the tip of the probe is generated by alternating current, while target tissue temperature ranges from 90° to 100°. The tissue is heated by radio frequency-induced intracellular oscillation of ionized molecules, which leads to elevation of intracellular temperature. The very high temperatures cause breakdown of tissue molecules. Inhomogeneous electric field with high voltage and selective current flow conduct to electric arcs (Figure 1).

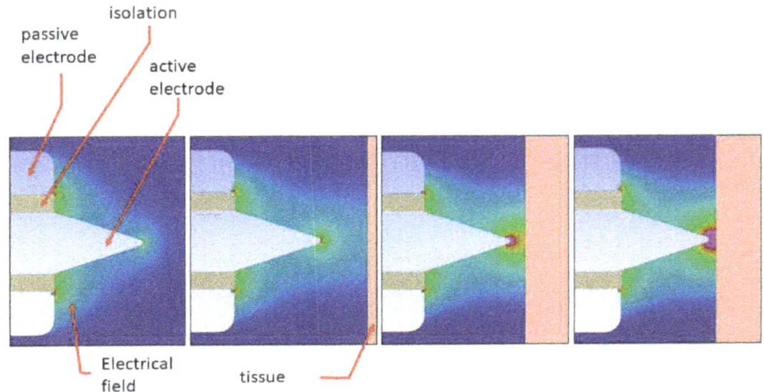

Figure 1. Inhomogeneous electric field, high voltage, and electric arc.

Bearing in mind that the vaporization or cutting process is the best accomplished with relatively low voltage, the coagulation is performed by arcing modulated high-voltage current to tissue. The electric arcs create cell bursts through the evaporation of cell content. By modulating the voltage, it is possible to cut and at the same time make locale coagulation. There is a galvanic separation between the power source and the surgery hand piece (Figure 2).

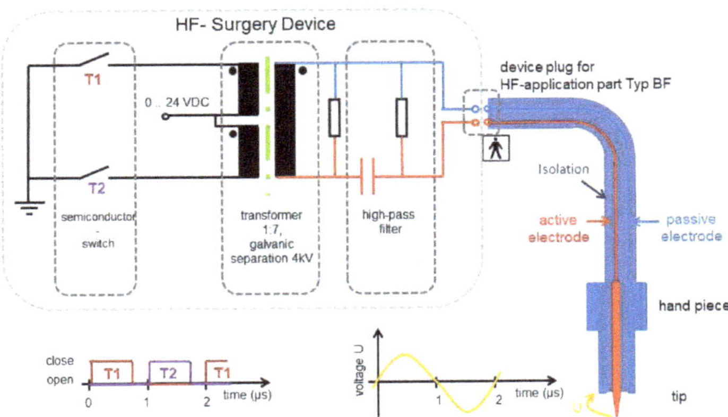

Figure 2. High-frequency (HF) generator.

First, the electrodes are placed away from the tissue, and then the gap between two electrodes is ionized. Due to ionization process, an electric arc discharge develops. In this approach, tissue burning is more superficial because the current is spread over the tissue area more than over the tip of electrode. The experimental set-up provides high-frequency power dissipation in the vicinity of the tip. This was demonstrated in the histological analysis by showing that there is more of a cutting effect in the tissue than coagulation, because there is no tissue nor cell denaturation in the cutting channel and surrounding tissue (Figure 3).

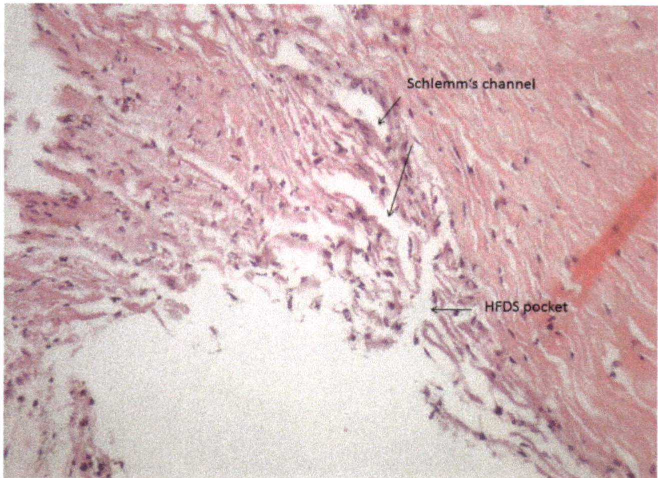

Figure 3. Histological analysis of a high-frequency deep sclerotomy (HFDS) cut (HE staining, human eye).

Two clear corneal incisions were used: one of 1.2-mm temporale or temporo-superior for introducing the abee tip and the other nasal with 0.8-mm created for cataract surgery. In the study, miochol (Acetylcholine chlorid 20 mg) was given intracamerally for the miosis. Anterior chamber was filled with a cohesive ophthalmic viscosurgical device (OVD). The standard high-frequency probe (Figure 4) as described above is consecutively inserted through the temporal paracentesis using a four mirror gonioscopic lens, until the probe is properly placed opposite nasally to the iridocorneal structure.

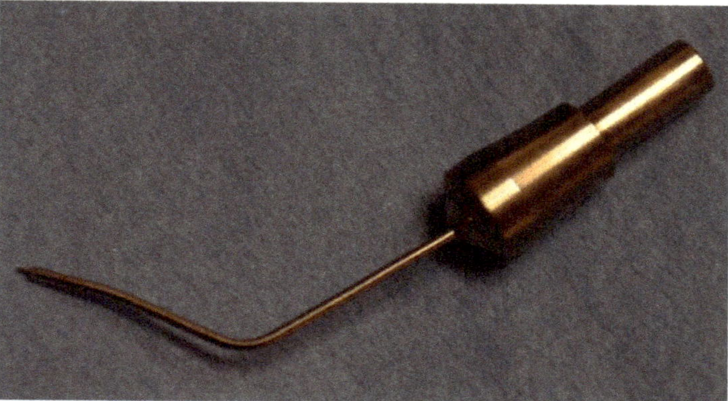

Figure 4. Abee® (HFDS) tip.

Then, six pockets were created consecutively in a row with approximately one tip length space between them. Recently, a new diathermic probe design was developed and used for all patients in the study. For all patients, six pockets were done. With the new abee tip design, three accesses to the anterior chamber are possible, i.e., temporal, superotemporal, or superonasal. The pockets can thus be placed nasally, nasally inferior, or nasally temporal. Consequent retreatments can be done easily. The external diameter is 0.9 mm. Oertli devices were used for the study, i.e., the Pharos and Catarhex 3, with the same setting.

The electric specifications of the probe remained unchanged. Pockets were created with the probe 1 mm through the trabecular meshwork, into Schlemm canal. The target was the insertion of a pocket of 1 mm into the sclera. The target deep sclerotomy pocket is approximatively 0.3 mm thick and 0.6 mm wide, resulting in a resorption surface area of 3.6 mm^2 (Figure 5a).

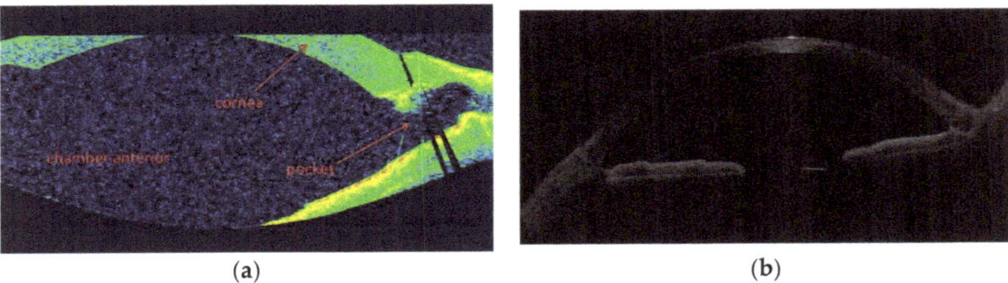

(a)   (b)

Figure 5. (a) OCT image of the anterior chamber: Overview after HFDS procedure with a newly created pocket in the chamber angle (Visante OCT, Zeiss, Oberkochen, Germany). (b) OCT image of the anterior chamber of a normal angle anterior before HFDS surgery.

For comparison, it is shown a normal anterior chamber and angle anterior (Figure 5b).

The tip's dimensions and ab interno approach make it compatible with the stipulations of minimally invasive glaucoma surgery.

Statistical analysis: statistical calculation was done with SPSS Program Version 22. Two-tailed student t-test was used for statistical evaluation of parametric data. Significance was set at a p value of <0.05.

3. Results

Mean age of patients with open-angle glaucoma was 76.8 ± 11.1 years (range: 35–88 years). 103 patients (50.2%) were female, and 102 patients (49.8%) male. One eye of each patient was included. During the whole period of 48 months, no repeat-surgery was needed in any included patient.

Mean preoperative IOP in the study population of 205 patients was 24.5 ± 4.3 mmHg (range 18 to 48 mmHg). Decimalised Snellen visual acuity (VA) increased from 0.46 ± 0.27 preoperative to 0.68 ± 0.27 postoperative. For all patients, the follow-up was 48 months. After 10 days, a slight increase of IOP was detected but was not statistically significant. Mean IOP after 48 months was 15.0 ± 1.7 mmHg (range 10 to 20 mmHg). The IOP drop was statistically significant ($p < 0.001$) at all measured postoperative intervals (Figure 6).

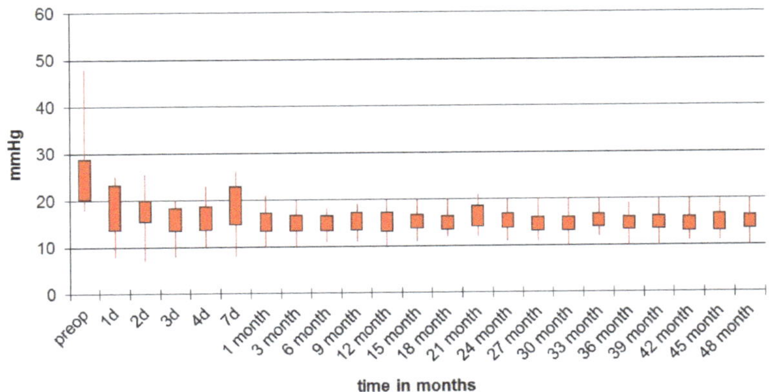

Figure 6. Intraocular pressure (IOP) follow-up of operated eyes during 48 months.

At month 48 after surgery, 54.7% of patients had an IOP < 15 mmHg, 77% had an IOP < 18 mmHg, and 84.9% had an IOP < 21 mmHg. Qualified success rate, defined as an IOP lower than 22 mmHg with medication, was 100% for all patients at 48 months (Figure 7).

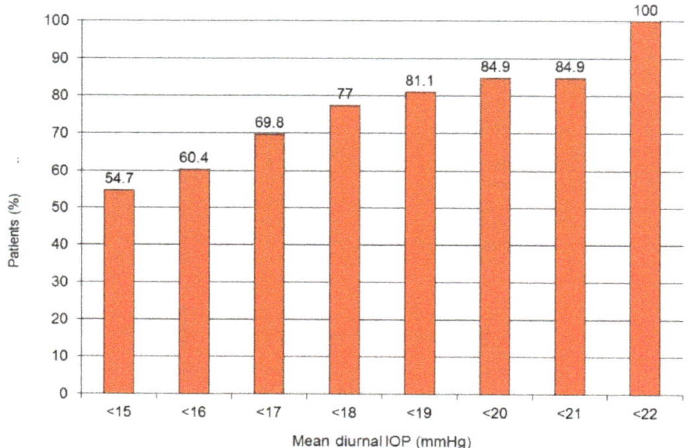

Figure 7. Qualified success rate after 48 months of follow-up.

The average preoperative administration of pressure-reducing number of antihypertensive eye agents was 2.6 ± 1.0. Following surgery, this value was decreased to 0.47 ± 0.59 after 1 month, 0.40 ± 0.54 after 3 months, 0.28 ± 0.62 after 6 months, 0.31 ± 0.49 after 12 months, 0.38 ± 0.57 after 24 months, 0.42 ± 0.83 after 36 months, and 0.48 ± 0.97 after 48 months (Figure 8).

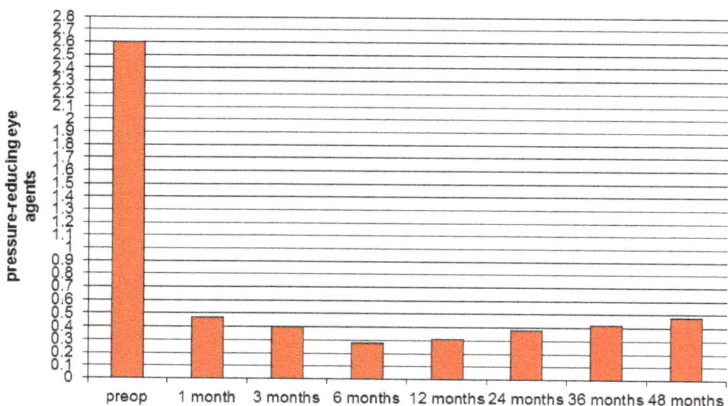

Figure 8. Progression of pressure-reducing eye agents pre-and postoperative during the 48 months, showing a marked and sustained postoperative reduction in numbers of agents needed.

There were no significant visual field changes (Octopus 900, Haag Streit, Switzerland) from baseline with a mean defect MD 7.06 ± 6.54 and loss of variance LV 26.4 ± 23.92. At 48 months, these values were MD 8.43 ± 2.11 and LV 24.1 ± 24.92 ($p = 0.22$ for MD, $p = 0.58$ for LV).

The mean excavation preoperative was 0.63 ± 0.22 and after 48 months 0.65 ± 0.21, which is not statistically significant ($p = 0.36$).

Temporary IOP elevation higher than 21 mmHg was observed in 18 of 205 eyes (8.7%). Four eyes (2%) showed transient fibrin formation. Fibrin was treated by topical dexamethasone and disappeared one day later.

4. Discussion

The present study has shown sustained IOP—lowering effect over 48 months using Schlemm's canal as an outflow pathway with a novel MIGS procedure. Looking into the past, the only possibility of glaucoma treatment was conservative medical management and more invasive glaucoma surgery. The advance MIGS procedure was intended to fill the existing gap regarding treatment. Early studies [19] have demonstrated its ability to lower IOP with minimal risk for mild to moderate glaucoma.

The HFDS ab interno method intends the creation of a direct channel between the anterior chamber and the Schlemm canal. Persistence of the sclerotomy has been investigated with a three-mirror goniolens 4 weeks after the procedures. The abee tip creates a deep sclerotomy with subsequent access of aqueous outflow to the scleral layer. Both aspects may facilitate a bypass effect of aqueous outflow. In an earlier study with deep sclerotomy ab interno, a significant IOP peak could be seen ten days postoperatively [17]. As shown in our study, with introduction of postoperative pilocarpine 2% eye drops application for the first 4 weeks, the high IOP peak amplitude could be avoided. This may have also been the case because the procedures were done combined with cataract surgery.

In general, MIGS procedures share five important features: ab interno approach, potential minimal trauma, ability to lower IOP, a high level of safety, and faster visual recovery. Hence, advantages of HFDS ab interno method, compared with trabeculectomy and perforating and nonperforating deep sclerectomy, seem to have a low rate of postoperative complications and a stable level of reduced IOP. Hypotension, a frequent finding in trabeculectomy, which is less frequently found in nonperforating deep sclerectomy, was not seen in the present population. The most frequent early complications in trabeculectomy are hyphaema (24.6%), shallow anterior chamber (23.9%), hypotony (24.3%), wound leak (17.8%), and choroidal detachment (14.1%). The most frequent late complications are iris incarceration (5.1%) and encapsulated bleb (3.4%) [21]. After HFDS, a transient IOP elevation was seen in 18 of 205 eyes (8.7%), on average occurring 10 days postoperatively. Four of 205 eyes (2%) had transient fibrin formation. Therefore, compared with penetrating or nonpenetrating techniques, HFDS seems to be a safe surgical technique [22–25].

An earlier study with deep sclerotomy ab interno had a complete success rate of 83% after 48 months [17,19] for open angle glaucoma. The present study shows that the success rate of 84.9% was similar. The slight improvement could be due to the fact that the procedures was carried out in combination with cataract surgery. It was particularly striking that the postoperative IOP range was significantly lower in the combined procedure. This effect could be explained by the additional reduction due to the cataract surgery, but also by the postoperative application of pilocarpine 2% during the first 4 weeks. If the postoperative IOP profile is analyzed, after 24 months there is a loss of the additional IOP-lowering effect from the cataract surgery will be apparent.

Advantages of HFDS include its comparative simplicity and short duration of the surgical procedure itself. Additionally, unlike ab externo filtering techniques, the technique avoids stimulation of episcleral and conjunctival tissues leading to fibroblast activation. Additionally, by comparing the histological alterations of HFDS with trabectome, another MIGS procedure, it has been shown that with trabectome tissue was damaged near the incision [26]. As trabectome was validated mainly in the pediatric population associated with long-term IOP control, recently a dual-blade device was developed for adult use but showing also smaller but still detectable damage to the tissue [27].

5. Conclusions

Although the number of six surgical pockets chosen was arbitrary, our results suggest that it may be sufficient to provide good long-term IOP-lowering efficacy and safety. With the introduction of the new abee-tip design, it is possible to perform surgical retreatment in the nasal and inferior quadrants of the trabecular meshwork. HFDS is a safe and minimally invasive method for glaucoma surgery with good long-term results. More studies are needed to confirm our findings in different populations and types of glaucoma.

Author Contributions: B.P. provided the treatment indication, developed the study design, acquired clinical data, and contributed to writing the paper. Z.C. contributed substantially to the development of the study design and substantially contributed to writing the paper. K.M. performed data analyses and substantially contributed advising regarding the paper contents. M.R. performed data anlysis and substantially contributed the study design. R.A. performed data analysis and contributed to writing the paper. All authors have read and agreed to the published version of the manuscript.

Funding: This research received no external funding.

Conflicts of Interest: The authors declare no conflict of interest.

References

1. Flaxman, S.R.; Bourne, R.R.A.; Resnikoff, S.; Ackland, P.; Braithwaite, T.; Cicinelli, M.V.; Das, A.; Jonas, J.B.; Keeffe, J.; Kempen, J.H.; et al. Global causes of blindness and distance vision impairment 1990–2020: A systematic review and meta-analysis. *Lancet Glob. Health* **2017**, *5*, 1221–1234. [CrossRef]
2. Goh, Y.W.; Ang, G.S.; Azuara-Blanco, A. Lifetime visual prognosis of patients with glaucoma. *Clin. Exp. Ophthalmol.* **2011**, *39*, 766–770. [CrossRef] [PubMed]
3. Razeghinejad, M.R.; Fudemberg, S.J.; Spaeth, G.L. The changing conceptual basis of trabeculectomy: A review of past and current surgical techniques. *Surv. Ophthalmol.* **2012**, *57*, 1–25. [CrossRef] [PubMed]
4. Vinod, K.; Gedde, S.J.; Feuer, W.J.; Panarelli, J.F.; Chang, T.C.; Chen, P.P.; Parrish, R.K. Practice preferences for glaucoma surgery: A survey of the American Glaucoma Society. *J. Glaucoma* **2017**, *26*, 687–693. [CrossRef] [PubMed]
5. Gedde, S.J.; Chen, P.P.; Heuer, D.K.; Singh, K.; Wright, M.M.; Feuer, W.J.; Schiffman, J.C.; Shi, W.; Primary Tube Versus Trabeculectomy Study Group. The Primary Tube Versus Trabeculectomy Study: Methodology of a Multicenter Randomized Clinical Trial Comparing Tube Shunt Surgery and Trabeculectomy with Mitomycin, C. *Ophthalmology* **2018**, *125*, 774–781. [CrossRef] [PubMed]
6. Vaziri, K.; Schwartz, S.G.; Kishor, K.S.; Fortun, J.A.; Moshfeghi, D.M.; Moshfeghi, A.A.; Flynn, H.W., Jr. Incidence of postoperative suprachoroidal hemorrhage after glaucoma filtration surgeries in the United States. *Clin. Ophthalmol.* **2014**, *9*, 579–584.
7. Takihara, Y.; Inatani, M.; Ogata-Iwao, M.; Kawai, M.; Inoue, T.; Iwao, K.; Tanihara, H. Trabeculectomy for open-angle glaucoma in phakic eyes vs in pseudophakic eyes after phacoemulsification: A prospective clinical cohort study. *JAMA Ophthalmol.* **2014**, *132*, 69–76. [CrossRef]
8. Jeganathan, V.S.; Ghosh, S.; Ruddle, J.B.; Gupta, V.; Coote, M.A.; Crowston, J.G. Risk factors for delayed suprachoroidal haemorrhage following glaucoma surgery. *Br. J. Ophthalmol.* **2008**, *92*, 1393–1396. [CrossRef]
9. Gedde, S.J.; Herndon, L.W.; Brandt, J.D.; Budenz, D.L.; Feuer, W.J.; Schiffman, J.C. Surgical complications in the Tube Versus Trabeculectomy Study during the first year of follow-up. *Am. J. Ophthalmol.* **2007**, *143*, 23–31. [CrossRef]
10. Johnson, D.H.; Johnson, M. Glaucoma surgery and aqueous outflow: How does nonpenetrating glaucoma surgery work? *Arch. Ophthalmol.* **2002**, *120*, 67–70. [CrossRef]
11. Schuman, J.S.; Chang, W.; Wang, N.; de Kater, A.W.; Allingham, R.R. Excimer laser effects on outflow facility and outflow pathway morphology. *Investig. Ophthalmol. Vis. Sci.* **1999**, *40*, 1676–1680.
12. Shaarawy, T. Glaucoma surgery: Taking the sub-conjunctival route. *Middle East Afr. J. Ophthalmol.* **2015**, *22*, 53–58. [CrossRef] [PubMed]
13. Maeda, M.; Watanabe, M.; Ichikawa, K. Evaluation of trabectome in open-angle glaucoma. *J. Glaucoma* **2013**, *22*, 205–208. [CrossRef] [PubMed]
14. Minckler, D.S.; Baerveldt, G.; Alfaro, M.R.; Francis, B.A. Clinical results with the Trabectome for treatment of open-angle glaucoma. *Ophthalmology* **2005**, *112*, 962–967. [CrossRef] [PubMed]
15. Resende, A.F.; Patel, N.S.; Waisbourd, M.; Katz, L.J. iStent® Trabecular Microbypass Stent: An Update. *J. Ophthalmol.* **2016**, *2016*, 2731856. [CrossRef]

16. Pfeiffer, N.; Garcia-Feijoo, J.; Martinez-de-la-Casa, J.M.; Larrosa, J.M.; Fea, A.; Lemij, H.; Gandolfi, S.; Schwenn, O.; Lorenz, K.; Samuelson, T.W. A Randomized Trial of a Schlemm's Canal Microstent with Phacoemulsification for Reducing Intraocular Pressure in Open-Angle Glaucoma. *Ophthalmology* **2015**, *122*, 1283–1293. [CrossRef]
17. Pajic, B.; Pajic-Eggspuehler, B.; Haefliger, I. New minimally invasive, deep sclerotomy ab interno surgical procedure for glaucoma, six years of follow-up. *J. Glaucoma* **2011**, *20*, 109–114. [CrossRef]
18. Abushanab, M.M.I.; El-Shiaty, A.; El-Beltagi, T.; Hassan Salah, S. The Efficacy and Safety of High-Frequency Deep Sclerotomy in Treatment of Chronic Open-Angle Glaucoma Patients. *BioMed Res. Int.* **2019**, *2019*, 1850141. [CrossRef]
19. Pillunat, L.E.; Erb, C.; Jünemann, A.G.; Kimmich, F. Micro-invasive glaucoma surgery (MIGS): A review of surgical procedures using stents. *Clin. Ophthalmol.* **2017**, *11*, 1583–1600. [CrossRef]
20. Edmunds, B.; Thompson, J.R.; Salmon, J.F.; Wormald, R.P. The National Survey of Trabeculectomy. III. Early and late complications. *Eye* **2002**, *16*, 297–303. [CrossRef]
21. Lindemann, F.; Plange, N.; Kuerten, D.; Schimitzek, H.; Koutsonas, A. Three-Year Follow-Up of Trabeculectomy with 5-Fluorouracil. *Ophthalmic Res.* **2017**, *58*, 74–80. [CrossRef] [PubMed]
22. Rodriguez-Una, I.; Rotchford, A.P.; King, A.J. Outcome of repeat trabeculectomies: Long-term follow-up. *Br. J. Ophthalmol.* **2017**, *101*, 1269–1274. [CrossRef] [PubMed]
23. Harju, M.; Suominen, S.; Allinen, P.; Vesti, E. Long-term results of deep sclerectomy in normal-tension glaucoma. *Acta Ophthalmol.* **2018**, *96*, 154–160. [CrossRef] [PubMed]
24. Bettin, P.; Di Matteo, F.; Rabiolo, A.; Fiori, M.; Ciampi, C.; Bandello, F. Deep Sclerectomy with Mitomycin C and Injectable Cross-linked Hyaluronic Acid Implant: Long-term Results. *J. Glaucoma* **2016**, *25*, 625–629. [CrossRef] [PubMed]
25. Pajic, B.; Pallas, G.; Gerding, H.; Böhnke, M. A novel technique of ab interno glaucoma surgery: Follow-up results after 24 months. *Graefes Arch. Clin. Exp. Ophthalmol.* **2006**, *244*, 22–27. [CrossRef] [PubMed]
26. Seibold, L.K.; Soohoo, J.R.; Ammar, D.A.; Kahook, M.Y. Preclinical investigation of ab interno trabeculectomy using a novel dual-blade device. *Am. J. Ophthalmol.* **2013**, *155*, 524–529. [CrossRef]
27. SooHoo, J.R.; Seibold, L.K.; Kahook, M.Y. Ab interno trabeculectomy in the adult patient. *Middle East Afr. J. Ophthalmol.* **2015**, *22*, 25–29. [CrossRef]

Article

Laser Vision Correction for Regular Myopia and Supracor Presbyopia: A Comparison Study

Bojan Pajic [1,2,3,4], Zeljka Cvejic [2], Horace Massa [3], Brigitte Pajic-Eggspuehler [1], Mirko Resan [4] and Harald P. Studer [1,5,*]

[1]   Eye Clinic Orasis, Swiss Eye Research Foundation, 5734 Reinach AG, Switzerland; bpajic@datacomm.ch (B.P.); brigitte.pajic@orasis.ch (B.P.-E.)
[2]   Faculty of Sciences, Department of Physics, University of Novi Sad, Trg Dositeja Obradovica 4, 21000 Novi Sad, Serbia; zeljka.cvejic@df.uns.ac.rs
[3]   Division of Ophthalmology, Department of Clinical Neurosciences, Geneva University Hospitals, 1205 Geneva, Switzerland; horace.massa@hcuge.ch
[4]   Faculty of Medicine of the Military Medical Academy, University of Defense, 11000 Belgrade, Serbia; resan.mirko@gmail.com
[5]   OCTLab, Dept. of Ophthalmology, University of Basel, 4056 Basel, Switzerland
*   Correspondence: harald.studer@gmail.com

Received: 29 December 2019; Accepted: 22 January 2020; Published: 27 January 2020

Abstract: A study to compare femto-presbyLASIK to standard myopia femto-LASIK refractive surgical correction with a total of 45 candidates was performed. The goal was to identify a more specific set of indications for presbyopia LASIK treatments. The results showed thoroughly good uncorrected visual acuity for myopia (decimal: 1.01 ± 0.15) as well as for presbyLASIK (decimal: 0.78 ± 0.17) corrections. Astigmatism was comparable in both groups and did not change significantly from preoperative ($0.98D \pm 0.53\,SD$) to postoperative ($1.01D \pm 0.50\,SD$). Our study results suggest, that presbyLASIK treatment is as safe and effective as regular LASIK myopia correction and can hence be recommended to treat presbyopia.

Keywords: supracor; myopia; presbyopia; presbyLASIK; refractive surgery; follow-up; clinical results; lasik; excimer laser

---

1. Introduction

Laser Assisted In-Situ Keratomileusis (LASIK) has become the most prevalent refractive surgical procedure for myopic and hyperopic corrections (about 1.2 million procedures a year in the US)—over the last two decades. Patient satisfaction is generally found to be very high between 92 and 98 percent [1,2]. Modern excimer laser systems can restore 20/20 uncorrected visual acuity (UCVA) in myopic eyes up to—10D and hyperopic up to +6D, feature tissue saving procedures [3], provide patient-specific, wavefront and/or topography guided ablation patterns, and can treat astigmatism with elliptically shaped patterns.

The concept of multifocal PresbyLASIK is a relatively new yet attractive surgical procedure for the correction of presbyopia. It involves two steps: (1) the correction of the ametropic state for distance vision, and (2) the multifocality addition for near vision. As most commonly reported in literature, multifocality is usually implemented by combining a central corneal curvature addition for near vision correction with a paracentral corneal curvature, adjusted to correct distance vision [4–9].

The downside of multifocal LASIK treatments is that they always represent a compromise between distance—and near vision correction, as they create unwanted aberrations—especially spherical aberrations in the central pupillary region. For this reason, modern PresbyLASIK treatment algorithms are wavefront-guided to minimize unwanted aberrations. Compared to treating presbyopia with the well-established multifocal intra ocular lens (IOL) implantation, PresbyLASIK has the advantage of being less invasive because the ocular globe does not need to be opened for treatment. However, the planning of PresbyLASIK procedures is much more demanding and requires experience in the interpretation of the preoperative corneal topography, wavefront analysis and the Zernike coefficients, which all have significant impact on the surgery result. Hence, it is not enough to simply apply the common LASIK indications to PresbyLASIK as well. Presumably, the lack of specific indications—along with missing to consider refraction and wavefront analysis parameters, might well be the reason for the not all encouraging LASIK presbyopia treatment results presented in literature [10,11].

In this study we present our PresbyLASIK results and compare them to a control group of regular LASIK myopia correction cases. Further, we propose an updated set of more specific indications for the PresbyLASIK treatment.

2. Materials and Methods

Our study population included two groups of eyes, 40 eyes of 20 study patients (female: 14, male: 6, age: 30 ± 9 years) were treated with regular myopic LASIK correction, and 50 eyes of 25 study patients (female: 14, male: 11, age: 52 ± 3.8 years) were treated for presbyopia with the PresbyLASIK Supracor treatment (B&L Technolas, Munich, Germany). All surgery procedures were performed by an experienced surgeon (BP) at the Eye Clinic ORASIS between January 2014 and February 2015.

The LDV femtosecond laser (Ziemer Ophthalmology, Port, Switzerland) was used to make the LASIK flaps, whereby the hinge was superior in all cases. The flap diameter depended on the corneal curvature and was between 8.5 and 9.0 mm. The Munnerlyn formula [12] along with the B&L-Technolas proprietary transition zone algorithm was used to calculate the ablation patterns for the regular myopia corrections. In a two-step approach a 2.0 mm laser spot diameter was first used to apply the basic correction, followed by the fine-tuning step with a 1.0 mm laser spot. An eye tracker system was active during all interventions. The cyclorotation is automatically compensated which is especially important for cases where astigmatism correction was included. Whenever the angle kappa was smaller or equal 6° the laser spot was placed in the pupil center, otherwise it was placed at the point of the arithmetic mean between the pupil center and the Purkinje reflex. After the laser correction, the flap was carefully placed back onto the stromal bed, making sure no visible wrinkles are left.

The same approach as for the myopia ablation patterns is also employed to calculate the Supracor PresbyLASIK ablation patterns, however, the target refraction is chosen such that the dominant eye was at 0.0D and the follow-eye was at −0.5D. The goal of a first application phase is to obtain a perfect mono-focal correction, and only in the second phase the central cornea is steepened by the multifocal curvature addition in the 3.0 to 6.0 mm zone. Considering that such an ablation could potentially induced spherical aberration, the Supracor procedure was performed wavefront-guided for all cases.

Corneal shape was assessed one week before and then 3 months after the LASIK treatment, with the Galilei G4 tomographer (Ziemer Ophthalmic Systems, Port, Switzerland). Thereby, each individual measurement was repeated three times in a row. So, only one measurement with the median astigmatism cylinder value is selected for further data evaluation.

The statistical methods used in this study were straightforward calculations of average and standard deviation, as described in any textbook of statistics. Excel software package (Microsoft, Redmond, WA, USA) was used to import and process data, as well as, to obtain the functional dependence of the selected parameters and their visualization. Two-tailed Student *t*-test was used for calculating the parametric datas and for the nonparametric test the Wilcoxon signed rank test was used. The level of significance was set at $p < 0.05$.

3. Results

All study data were carefully analyzed and are presented in the following. Manifest refraction of the whole study population was sph: −1.99D ± 3.48 (SD), cyl: −0.80D ± 0.79 (SD) before, and sph: −0.03D ± 0.27 (SD), cyl: −0.18D ± 0.32 (SD) after the surgery. While the myopia group had a manifest refraction of sph: −3.53D ± 2.48 (SD), cyl: −0.93D ± 0.84 (SD) before, and sph: +0.05D ± 0.19 (SD), cyl: −0.09D ± 0.27 (SD) after the surgery, the presbyopia group had a manifest refraction of sph: −1.02D ± 3.78 (SD), cyl:−0.72D ± 0.75 (SD) before, and sph: −0.09D ± 0.31 (SD), cyl: −0.24D ± 0.34 (SD) after the surgery.

3.1. Uncorrected Visual Acuity

Uncorrected visual acuity (UCVA) was assessed before (decimal scale: 0.31 ± 0.24) and after the treatment (decimal scale: 0.88 ± 0.20) for the whole study population. In the myopia group UCVA was 0.19 ± 0.16 preoperatively, and 1.01 ± 0.15 postoperatively. In the presbyopia group UCVA was 0.38 ± 0.25 preoperatively, and 0.78 ± 0.17 postoperatively. A statistically significant improvement was observed in both groups ($p < 0.0001$). In 4 eyes (4.3%, two in myopia and two in presbyopia group), UCVA remained unchanged, and for two eyes (2.2%, all in the presbyopia group) UCVA was slightly worse than before the treatment. For all other eyes (93.5%), UCVA improved (see Table 1). The failure to achieve the target refraction and UCVA could be corrected with a re-treatment.

Table 1. Uncorrected visual acuity after regular myopia and Supracor presbyopia treatment.

	Average		Minimum		Maximum	
	Decimal	**Metric**	**Decimal**	**Metric**	**Decimal**	**Metric**
All eyes	0.88 ± 0.20	5.3/6	0.32	1.9/6	1.25	7.5/6
Myopia	1.01 ± 0.16	6.1/6	0.63	3.8/6	1.25	7.5/6
Supracor	0.78 ± 0.17	4.7/6	0.32	1.9/6	1.00	6/6

While the multifocal presbyopia treatments showed a small trend ($R^2 = 0.01$) towards greater corrections having a lower postoperative UCVA, this trend could not be observed ($R^2 = 1 \times 10^{-5}$) for the myopic corrections (see Figure 1).

3.2. Residual Stromal Bed and Percentage Tissue Altered

The average residual stromal bed thickness for all cases was 385 µm ± 49 (SD), with a minimum of 285 µm, and a maximum of 510 µm. The recently published and already widely accepted new criterion for LASIK safety, the percentage tissue altered (PTA) value [13] was 32.0% ± 3.0 (SD), with a minimum of 22% and a maximum of 47%.

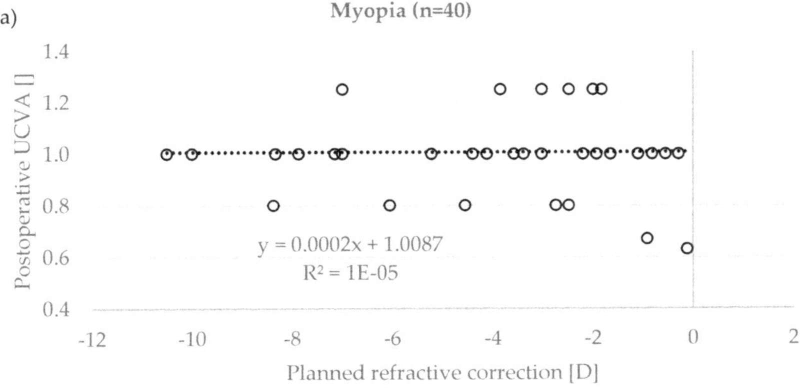

Figure 1. *Cont.*

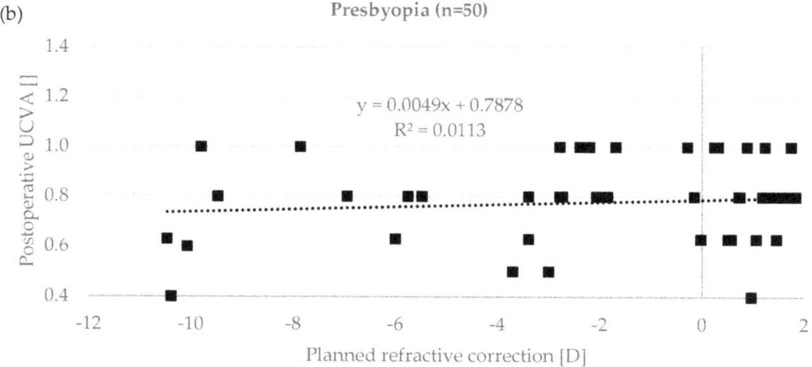

Figure 1. Trend analysis for presbyopia treatments (**b**) showed a mild correlation ($R^2 = 0.01$ with slope 0.005) that greater dioptric corrections resulted in lower uncorrected visual acuity (UCVA). No such correlation ($R^2 = 1E\text{-}05$ with slope < 0.001) between dioptric correction and resulting UCVA could be observed in the myopia (**a**) results, though.

3.3. Astigmatism Change

Corneal astigmatism, assessed with the Galilei G4, showed very little difference ($p > 0.6$), between preoperative (0.98D ± 0.53 SD) and postoperative measurements (1.01D ± 0.50 SD) over the whole study population. Likewise, the myopia group had preoperative astigmatism of 1.17D ± 0.63 (SD), which changed to postoperative astigmatism of 0.78D ± 0.74 (SD) $p > 0.02$. The presbyopia group had preoperative astigmatism of 1.02D ± 0.42 (SD), which changed to postoperative astigmatism of 0.96D ± 0.58 (SD) $p > 0.02$. Also, Figure 2 shows double-angle polar plots of pre-, and post-operative astigmatism measurements for both groups, myopia and presbyopia, and suggests only small changes in astigmatism from pre—to postoperative measurements. However, the Alpins method [14] vector analysis (including the astigmatism orientation) showed that astigmatism changed significantly, by 0.78D ± 0.74 (SD) for myopia and by 0.96D ± 0.58 (SD) for presbyopia, respectively. The intended astigmatic change (planned astigmatic LASIK ablations) was 0.94D ± 0.84 (SD) for myopia group and 0.65D ± 0.41 (SD) for presbyopia group.

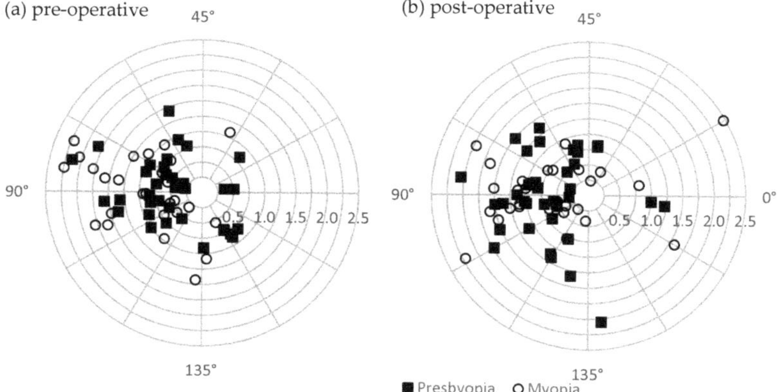

Figure 2. Double-angle polar plots comparing (**a**) pre- to (**b**) postoperative astigmatism. Myopia results are shown as circles and presbyopia results are shown as solid squares. No obvious differences between myopia and presbyopia correction were visible.

3.4. Corneal Curvature Change

The planned shape change of the anterior corneal surface was compared to the measured corneal surface average corneal curvature–AvgK (postoperative Galilei minus preoperative Galilei). In an ideally case, in Figure 3, all data points would lie on a diagonal line ($y = 1.0x + 0.0$, with an $R^2 = 1.0$). The top plot (b) in Figure 3 shows that the above graph (b) in Figure 3 shows that the presbyopia corrections showed a better match between the planned and achieved spherical correction ($y = 0.67x - 0.09$, $R^2 = 0.93$) compared to the myopia correction (see graph (a) in Figure 3) ($y = 0.45x - 1.22$, $R^2 = 0.47$).

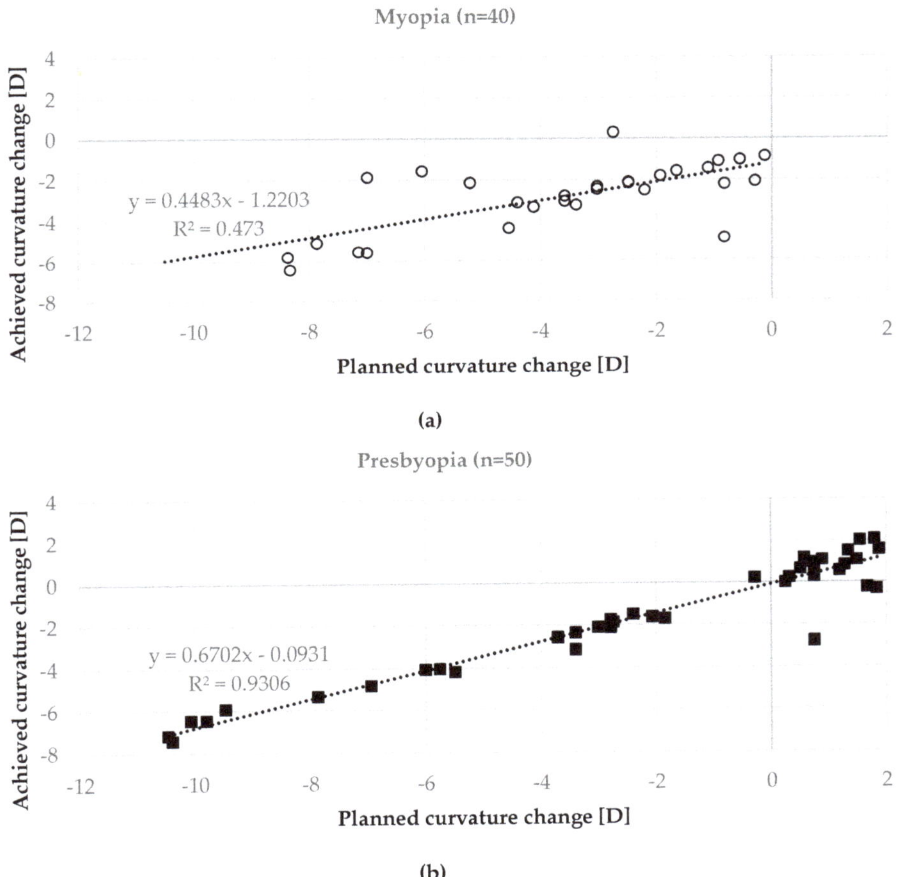

Figure 3. Comparing planned to achieved shape change of anterior corneal surface. Presbyopia treatments (**b**) showed a better match ($R^2 = 0.93$ and slope 0.67) than (**a**) myopia ($R^2 = 0.45$ and slope 0.47).

4. Discussion

In this study, femto-LASIK refractive surgical treatments were performed in two groups of study populations. While one of the groups received a standard myopia treatment, the second group was treated with a specific multifocal presbyLASIK algorithm to treat presbyopia. Study data, such as uncorrected visual acuity (UCVA), were carefully analyzed and presented as study results.

In almost all the cases (>93%), UCVA was improved after the surgery, with respect to the baseline assessment before the treatment. Thereby, the standard femto-LASIK myopia treatment showed slightly better results than the presbyLASIK treatment, where outcome UCVA was lower. This most likely arises from the fact that the subjective tolerance to eye surgical treatments including multifocal optics is greater than in monofocal treatments. Even a small deviation of 0.5D from the target value can be disturbing for the patient. Accordingly, there was a small but significant trend in recent results, where higher correction of presbyLASIK lead to lower postoperative UCVA. No similar trend was observed in the standard myopia treatment cases, where even high corrections of up to −10.5D had perfect vision with UCVA result of decimal 1.0.

The completely positive UCVA outcome was achieved, although the results suggest that astigmatism was drastically undertreated. Even though the laser was programmed with astigmatism correcting ablation patterns for all cases with treatable preoperative astigmatism, significant postoperative corneal astigmatism was observed in the topography measurements of both study groups. This observation is not supported by estimates of apparent refraction, where the cylinder of astigmatism decreased from −0.80D ± 0.79 (SD) to −0.18 D ± 0.33 (SD) ($p < 0.00001$). Thus, it could be assumed that the topography measurements were not accurate enough, but we did not have additional data to examine this in more details.

After surgical treatment, the mechanical strength of the cornea depends only on the residual stroma. It is well known that when the residual stroma is too thin to handle IOP, both its anterior and posterior surface will bulge forward as in ectasia. Nowadays, residual stroma thickness (RST) is accepted to be at least 250 μm. Conventionally, thickness less than 480 is considered abnormal, while some surgeons consider preoperative corneal thickness less than 500 μm unsuitable for LASIK. While planning the treatments, 250 μm of the RST was suggested as it accounts for 50% of 500 μm, which is the minimum corneal thickness recommended for LASIK. Also, high percentage of tissue altered (PTA) is the most common cause of ectasia after LASIK in patients with normal preoperative corneal topography. According to Santiago et al. [13], high values of PTA, especially higher than 40% is important factor in the development post-Lasik ectasia with normal preoperative corneal topography. For that reasonPTA should be taken as one of the screening parameter for refractive surgery candidates. In addition to this criterion, we have aimed at a postoperative thickness of 300 μm residual cornea stroma or more for all cases to allow re-treatment if necessary. None of the eyes in this study developed postoperative ectasia.

The observation, that postoperative central corneal curvature was steeper than planned comes from the fact that the algorithm used for the multifocality add-on causes a central steepening within the 3.0 mm zone. This is because the central zone is intended to be used for near vision. The 3.0 to 6.0 mm zone is adjusted to correct intermediate and distance vision. Both eyes thereby are treated the same way, except that the non-dominant eye is targeted at −0.5D from plano.

The results in this study suggest that the used procedure for presbyLASIK improves vision of presbyopia patients. Hence, we propose to include subjects for the presbyopia treatment older than 48 years, with a mesopic pupil of 3.5–6.5 mm diameter, and with a postoperative Kmean between 35.0D and 47.0D. The central near vision add-ons should be chosen >2.0D. We further suggest that cases where eye surgery was previously performed should not be included, especially cases where corneal refractive surgery was performed. Also, those cases with irregular corneal morphology, lens opacification, natural preoperative multifocality in the cornea, or a central corneal thickness CCT < 500 μm microns should not be included.

Author Contributions: B.P. and the team from the Orasis clinic in Switzerland conducted the clinical study and collected all the data and contributed substantially to writing the manuscript. Z.C. contributed substantially to the methodology, collecting datas and contributed substantially to writing the manuscript. H.M. performed data analyses and substantially contributed advising regarding the paper contents. B.P.-E. perform substantially data curation and contributed the study design. M.R. performed data anlysis and substantially contributed the study design. H.P.S. did the data analysis, statistical analysis and created all the plots and figures, and contributed substantially to writing the manuscript. All authors have read and agreed to the published version of the manuscript.

Funding: The clinical study funded by the Swiss research project CTI 13404.1, by the commission for technology and innovation, Switzerland.

Conflicts of Interest: All authors declare to have no conflict of interest.

References

1. Tahzib, N.G.; Bootsma, S.J.; Eggink, F.A.; Nabar, V.A.; Nujits, R.M. Functional outcomes and patient satisfaction after laser in situ keratomileusis for correction of myopia. *J. Cataract Refract. Surg.* **2005**, *10*, 1943–1951. [CrossRef] [PubMed]
2. Bailey, M.D.; Mitchell, L.G.; Zadnik, K. Reasons for recommending laser in situ keratomileusis. *Optom. Vis. Sci.* **2005**, *79*, 10. [CrossRef]
3. McGhee, C.N.; Graig, J.P.; Sachdev, N.; Weed, K.H.; Brown, A.D. Functional, psychological, and satisfaction outcomes of laser in situ keratomileusis for high myopia. *J. Cataract Refract. Surg.* **2000**, *26*, 497–509. [CrossRef]
4. Alarcón, A.; Anera, R.G.; Villa, C.; Jiménez del Barco, L.; Gutierrez, R. Visual quality after monovision correction by laser in situ keratomileusis in presbyopic patients. *J. Cataract Refract. Surg.* **2011**, *37*, 1629–1635. [CrossRef] [PubMed]
5. Uthoff, D.; Pölzl, M.; Hepper, D.; Holland, D. A new method of cornea modulation with excimer laser for simultaneous correction of presbyopia and ametropia. *Graefe's Arch. Clin. Exp. Ophthalmol.* **2012**, *250*, 1649–1661. [CrossRef] [PubMed]
6. Baudu, P.; Penin, F.; Arba Mosquera, S. Uncorrected binocular performance after biaspheric ablation profile for presbyopic corneal treatment using AMARIS with the PresbyMAX module. *Am. J. Ophthalmol.* **2013**, *155*, 636–647. [CrossRef] [PubMed]
7. Jackson, W.B.; Tuan, K.M.; Mintsioulis, G. Aspheric wavefrontguided LASIK to treat hyperopic presbyopia: 12-month results with the VISX platform. *J. Refract. Surg.* **2011**, *27*, 519–529. [CrossRef] [PubMed]
8. Jung, S.W.; Kim, M.J.; Park, S.H.; Joo, C.-K. Multifocal corneal ablation for hyperopic presbyopes. *J. Refract. Surg.* **2008**, *24*, 903–910. [CrossRef] [PubMed]
9. Luger, M.H.; Ewering, T.; Arba-Mosquera, S. One-year experience in presbyopia correction with biaspheric multifocal central presbyopia laser in situ keratomileusis. *Cornea* **2013**, *32*, 644–652. [CrossRef] [PubMed]
10. Pajic, B.; Pajic-Eggspuehler, B.; Mueller, J.; Cvejic, Z.; Studer, H. A Novel Laser Refractive Surgical Treatment for Presbyopia: Optics-Based Customization for Improved Clinical Outcome. *Sensors* **2017**, *17*, 1367. [CrossRef] [PubMed]
11. Pajic, B.; Massa, H.; Eskina, E. Presbyopiekorrektur mittels Lasechirurgie. *Klin. Monatsbl. Augenheilkd.* **2017**. [CrossRef]
12. Munnerlyn, C.R.; Koons, S.J.; Marshall, J. Photorefractive keratectomy: A technique for laser refractive surgery. *J. Cataract Refract. Surg.* **1988**, *14*, 46–52. [CrossRef]
13. Santhiago, M.R.; Smadja, D.; Gomes, B.F.; Mello, G.R.; Monteiro, M.R.L.; Wilson, S.E.; Randelman, J.B. Association Between the Percentage Tissue Altered and Post-Laser In Situ Keratomileusis Ectasia in Eyes With Normal Preoperative Topography. *Am. J. Ophthalmol.* **2014**, *158*, 87–95. [CrossRef]
14. Alpins, N.A. A new method of analyzing vectors for changes in astigmatism. *J. Cataract Refract. Surg.* **1993**, *19*, 524–533. [CrossRef]

Article

A Feasibility Study of a Novel Piezo MEMS Tweezer for Soft Materials Characterization

Fabio Botta *, Andrea Rossi and Nicola Pio Belfiore

Dipartimento di Ingegneria, Università degli Studi Roma Tre, Via della Vasca Navale, 79, 00146 Roma, Italy; andrea.rossi@uniroma3.it (A.R.); nicolapio.belfiore@uniroma3.it (N.P.B.)
* Correspondence: fabio.botta@uniroma3.it; Tel.: +39-06-57333491

Received: 1 May 2019; Accepted: 28 May 2019; Published: 2 June 2019

Abstract: The opportunity to know the status of a soft tissue (ST) in situ can be very useful for microsurgery or early diagnosis. Since normal and diseased tissues have different mechanical characteristics, many systems have been developed to carry out such measurements locally. Among them, MEMS tweezers are very relevant for their efficiency and relative simplicity compared to the other systems. In this paper a novel piezoelectric MEMS tweezer for soft materials analysis and characterization is presented. A theoretical approach has developed in order to carry out the values of the stiffness, the equivalent Young's modulus, and the viscous damping coefficients of the analyzed samples. The method has been validated by using both Finite Element Analysis and data from the literature.

Keywords: MEMS; tweezer; piezoelectric; soft tissue; microsurgery

---

1. Introduction

Early diagnosis is crucial to prevent disease progressions [1]. The correlation between the mechanical characteristics of a soft tissue and its status has been highlighted by several studies in many cell types like cancer cells, epithelial cells and laminopathies associated with diseases of the nuclear membrane [2]. In cancer cells, for example, the malignant transformation influences the mechanical properties by disruption and/or reorganization of cytoskeleton [3]. The stiffness, the elastic modulus, and the dynamic viscosity are among the most commonly used mechanical quantities to evaluate the state of a soft tissue. However the possibility to identify a diseased soft tissue in situ, at the microscale, can increase significantly the possibility of the cure.

Different techniques are used to perform such measurements as micropipette aspiration [4], magnetic bead twisting [5], atomic force microscopy [6], optical tweezers [7] or mechanical tweezers [8]. In recent years, researchers focused on the developing of micro-electro-mechanical systems (MEMS) tweezers because of their efficiency and relative simplicity with respect to other systems [9]. New gripping tweezers have been built taking cue from the kinematics of articulated mechanisms [10] as the property of parallelograms [11–15]. More complex motion, with an increasing of the number of degrees of freedom, can be obtained adopting the concept of lumped compliance. In fact, concentrated flexibility has been used in flexure hinges in several investigations [16–20]. In order to improve the displacement accuracy and to lower the stress levels in the flexible elements a new conjugate surfaces flexure hinge (CSFH) flexure has been proposed [21–25] and optimized [26,27]. The system has shown promising results in terms of of versatility and applicability to different types of tissue [28]. Thanks to their high level of miniaturization, MEMS-Technology based tweezers can be employed also in minimal invasive [29–31] or gastrointestinal surgery [32], and more generally speaking, in endoluminal surgery, for example, in TEM [33–35].

The actuation of the tweezer can be obtained by different systems, such as linear electrostatic actuators [36,37], rotary electrostatic actuators [38,39] or electrothermal actuation [40–46]. However the advent of smart materials in the last decades has greatly increased the possibilities of development of new smart structures [47–49] with the ability, not possible for the traditional systems, to adapt to external conditions variations. Because of their speed of response, low power consumption and high operating bandwidth the piezoelectric materials are, among the smart materials, the most promising ones for active vibration control [50–53] and MEMS applications [54]. However the research in the latter field mainly focuses on the energy harvesting [55–58] or MEMS tweezers for manipulating micro-objects and microassembly [59–61].

This paper presents the design of a novel piezoMEMS tweezer for the analysis and characterization of soft materials. Each jaw can be built as a sandwich composite beam and activated by an electric field.

The tweezer is supposed to be force controlled because the piezoelectric materials produce a stress proportional to the applied electrical field. Furthermore, a sensor is supposed to be integrated into the structure for displacement control. This gives rise to improve the analysis of the cell properties. In fact, the displacement-controlled actuator is able to identify the beginning of a rupture during straining or softening, while a force-controlled system maintains a constant force regardless of the required displacement [62]. The action of the piezoelectric material has been modeled by the Pin Force Model [63].

By applying symmetric electric fields to the composite jaws, they will bend in opposite directions allowing the gripping of the soft tissue.

A new mathematical model to measure the stiffness, the equivalent Young's modulus, and the viscous damping coefficient of the soft tissue has been developed. The model has been tested on three different soft test specimens and the results were in good agreement with those obtained by COMSOL finite elements code.

2. The Adopted Piezo-Mems Microgripper

The purpose of this paper is to develop a theoretical model of a piezo-MEMS microgripper, which can be used to characterize a grasped sample tissue. Therefore, the actual fabrication process of this instrument will not be herein considered. However, for the sake of completeness, a selection of possible materials and technological processes is briefly recalled.

In the last decades, several actuation methods have been proposed to induce motion in MEMS devices such as electrostatic, thermal and piezoelectric. The electrostatic devices are widely adopted but the piezoelectric MEMS's offer some attractive advantages: lower power consumption, broader bandwidth and approximately ten times lower voltages to obtain the same given displacement [64]. Furthermore the piezoelectric materials can be manufactured using the same MEMS conventional technologies and for this reason these materials have been preferred to develop piezo-MEMS devices in the last decades. Typical applications include vibration energy harvesters [65,66], resonators [67], capacitors [68], micro sensors/actuators [69,70], micromachined ultrasonic transducer [71], gyroscopic sensors [72], microlens [73,74], 1D and 2D micro-scanners [75,76].

The piezoelectric materials can be gathered into two groups: ferroelectric (lead zirconate titanate, PZT compounds) and non-ferroelectric (ZnO and AlN). The piezoelectric characteristics (piezoelectric coefficient, Q factor, dielectric constant, etc.) rely on the crystalline structure. In fact ZnO and AlN thin-films show wurtzite structure that entails lower piezoelectricity if compared to PZT materials (perovskite structure). Nevertheless ZnO and AlN exhibit large mechanical stiffness, high Q factor and do not require a polarization process so they can be attractive for sensors applications. PZT thin-films provide high piezoelectric properties, lower cost and stability against temperature but require a poling process before using the piezo-MEMS device [77]. The electric field poling direction depends on the functional configuration of the piezo-MEMS.

Usually the piezoelectric MEMS actuators/sensors are based on cantilever structures and the number of active layers identifies their working configuration:

- unimorph (one piezoelectric layer coupled with an inactive structural layer) or
- bimorph (a structural layer sandwiched by two active layers).

When a PZT bimorph bending beam and the transverse piezoelectric effect (d_{31}) are considered, the PZT layers must be poled in opposite directions in order to maximize the bending action. Then the electrodes of the PZT layers in contact with the structural layer share the same electric potential, whereas the outer electrodes share the same opposite sign potential (see Section 3). The design and fabrication process of piezo-MEMS devices have been extensively explored for the above mentioned unimorph and bimorph cantilever configurations using the aforementioned materials:

- AlN [67,68,71];
- ZnO [70]; and
- PZT [66,69,75,78].

Various technologies can be applied to deposit piezoelectric thin-films such as

- pulsed laser deposition (PLD);
- chemical vapour deposition (CVD);
- screen printing;
- sol-gel and
- radio frequency sputtering.

Usually the sol-gel and sputtering methods are the preferred ones both for research and commercial production because they allow the piezoelectric thin-film to be homogeneously deposited on large Si wafers [79]. The beam structural layer can be metal-based or silicon-based and the etching processes could be accomplished relying on conventional techniques (RIE, D-RIE, ECR).

A schematic view of the target piezo-MEMS tweezer is qualitatively depicted in Figure 1. The system can be obtained by using a multilayer wafer that can be built by using the above mentioned techniques. At the end of the process, the whole microsystem is composed of two bimorph beams (a)-(b)-(c) (see Figure 1). The two bimorph beams are supported by the handle layer (d). The specific steps of the process (deposition, etching, exposure, etc.) will depend on the peculiar materials and technology selected for the construction. In the case under study, the system is conceived in such a way that the mask geometry could be quite elementary for any etching step.

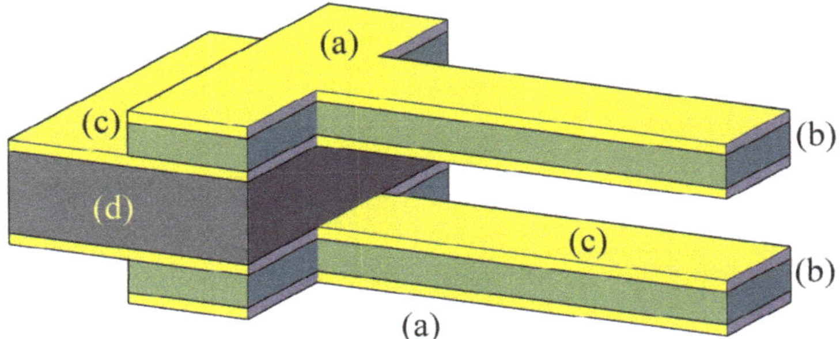

Figure 1. A schematic view of the microsystem: outer layers with a same voltage V_0 (a); structural material for the beams (b); internal layers with voltage V_i (c) and structural layer for the handle (d).

Another possible layout is represented in Figure 2, where two new layers (e) have been added with respect to the previous example of fabrication. Layers (a), (b), (c), and (d) have the same function as described for the previous layout. The second design is better for the operational aspects because a clamping tooth has been added for each jaw. However, these two teeth are more difficult to obtain during the process because they require at least two more deposition layers (e) and also a more complex series of intermediate etching-deposition steps that depend on the selected materials.

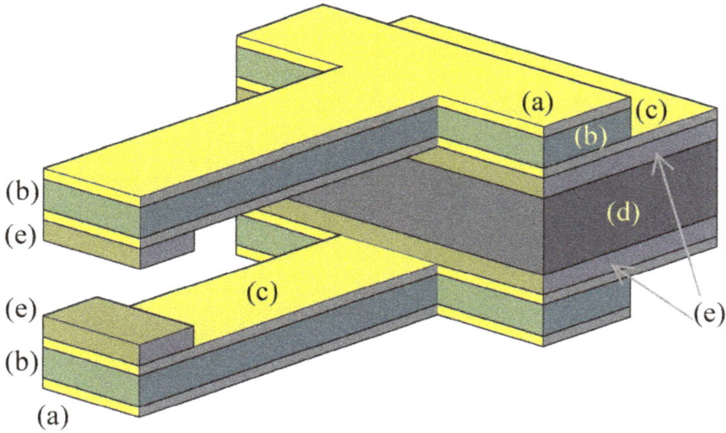

Figure 2. An alternative layout for the microgripper: outer layers with a same voltage V_0 (**a**); structural material for the beams (**b**); internal layers with voltage V_i (**c**) and structural layer for the handle (**d**).

3. Modeling Piezoelectric Actions on the Microbeams

Concerning the piezoelectric actions, many studies [63,80–82] have confirmed that, under certain assumptions (piezoelectric plates are perfectly bonded to the structure, the ratio between their thickness and the thickness of the beam is very low and their inertia and mass negligible with respect to those of the beam) the stresses applied to the beam can be considered as they were concentrated at the piezoelectric plate ends. Moreover, if the electric field to the upper plate is opposite in sign to the one applied to the lower plate, such action is equivalent to a flexural moment (see Figure 3) applied to the end of the beam of magnitude equal to:

$$M_a(t) = \frac{\psi}{6 + \psi} E_a c T_a T_b \Lambda(t) \tag{1}$$

with

$$\begin{cases} \Lambda(t) = \frac{d_{31}}{T_a} V(t) \\[2mm] \psi = \frac{E_b T_b}{E_a T_a} \end{cases} \tag{2}$$

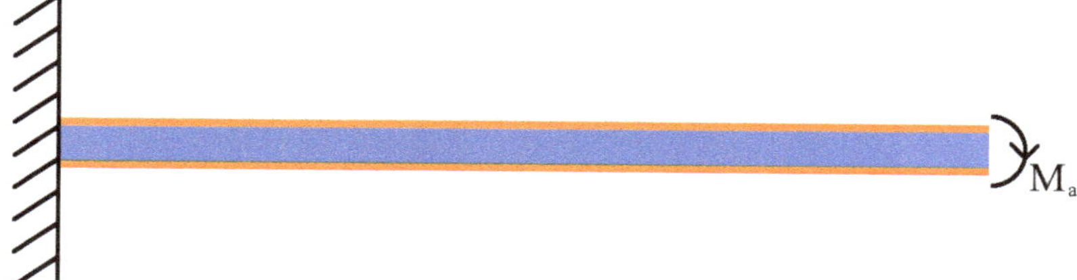

Figure 3. Piezoelectric action (PIN force model).

By activating the piezoelectric plates pairs on the two jaws-beams of the tweezer in opposite manner the two moments M_a will have opposite sign and the beams will bend in opposite direction allowing the gripper to grasp the soft tissue, as depicted in Figure 4.

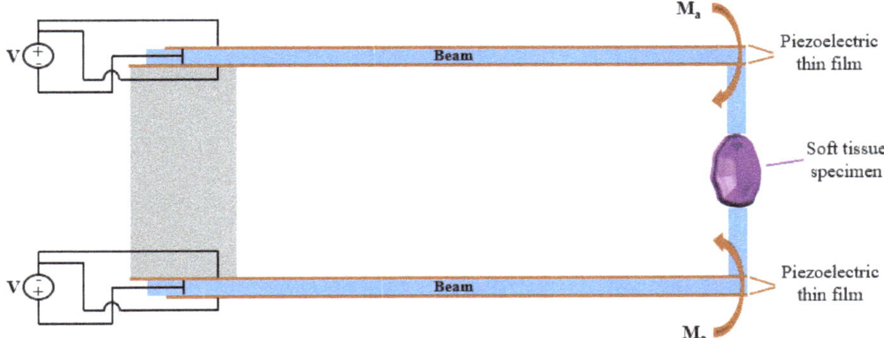

Figure 4. PiezoMEMS grip of the soft tissue.

3.1. Static Model

The value of the stiffness K_{ST} and of the equivalent Young's modulus E_{ST} can be established by a static test with a constant voltage V_0 applied to the piezoelectric plates: $V(t) = V_0$. In this case, the bending of the beams will cause the compression of the soft tissue and consequently a reaction force F_{ST} will be applied from the soft tissue to the beams, through the clamp teeth (see Figure 5).

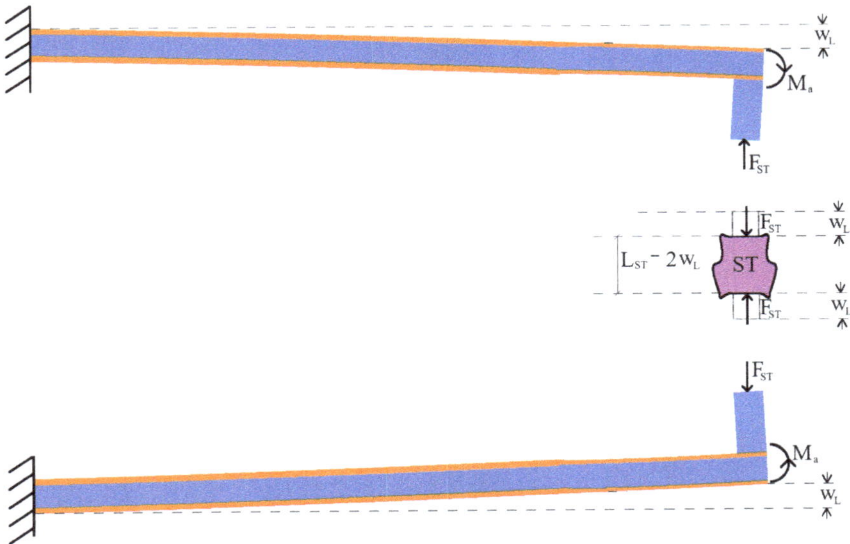

Figure 5. Scheme for the calculation of the tip displacement.

Considering that the Euler Bernoulli model is appropriate for the present case because of the high length-to-thickness aspect ratio of the beams, the tip deflection:

$$w_L = \frac{M_a L^2}{2 E_b I_b} - \frac{F_{ST} L^3}{3 E_b I_b}, \tag{3}$$

is due to a combination of the loads F_{ST} and M_a. Therefore, the first load

$$F_{ST} = \frac{3E_b I_b}{L^3}\left(\frac{M_a L^2}{2E_b I_b} - w_L\right) \tag{4}$$

can be obtained by measuring w_L.

Due to the high value of the axial stiffness of the clamp teeth, the deflection w_L can be considered equal to the axial displacement of the soft tissue (see Figure 5) so that the axial strain

$$\epsilon_{ST} = \frac{2w_L}{L_{ST}} \tag{5}$$

is calculated.

Under the assumption of linear elastic behavior for the soft tissue (ST), the relationship between the normal stress σ_{ST} applied to the cross section of the soft tissue and ϵ_{ST} can be written as:

$$\sigma_{ST} = \frac{F_{ST}}{S_{ST}} = E_{ST}\frac{2w_L}{L_{ST}} \tag{6}$$

from which the equivalent Young's modulus

$$E_{ST} = \frac{F_{ST}L_{ST}}{2S_{ST}w_L} \tag{7}$$

is evaluated. Moreover, the stiffness

$$K_{ST} = \frac{2E_{ST}S_{ST}}{L_{ST}} \tag{8}$$

is calculated given the dimensions of the ST sample.

3.2. Dynamics Model

To determine the value of the viscous damping C_{ST}, a dynamic test has been conceived by applying a variable harmonic excitation voltage $V(t) = V_0 sin(\omega t)$.

By referring to δL_{int}, δL_{in}, δL_a and δL_{ST} as the virtual works of internal, inertial, actuator and ST forces, respectively, and to δL_{in_m} as the virtual work of the inertial forces of the clamp teeth, the virtual work principle can be written as:

$$\delta L_{int} = \delta L_{in} + \delta L_a + \delta L_{in_m} + \delta L_{ST} \tag{9}$$

where (the quantities over signed by a tilde are virtual quantities):

$$\begin{cases} \delta L_{int} = E_b I_b \int_0^L \frac{\partial^2 w}{\partial x^2}\frac{\partial^2 \tilde{w}}{\partial x^2}dx \\[2mm] \delta L_{in} = -\rho S \int_0^L \frac{\partial^2 w}{\partial t^2}\,\tilde{w}\,dx \\[2mm] \delta L_a = M_a \frac{\partial \tilde{w}}{\partial x}\Big|_{x=L} \\[2mm] \delta L_{in_m} = -m\frac{\partial^2 w_L}{\partial t^2}\,\tilde{w}_L \\[2mm] \delta L_{ST} = F_{ST}\,\tilde{w}_L \end{cases} \tag{10}$$

with (see Figure 6):

$$F_{ST} = K_{ST}w_L + C_{ST}\,\dot{w}_L \tag{11}$$

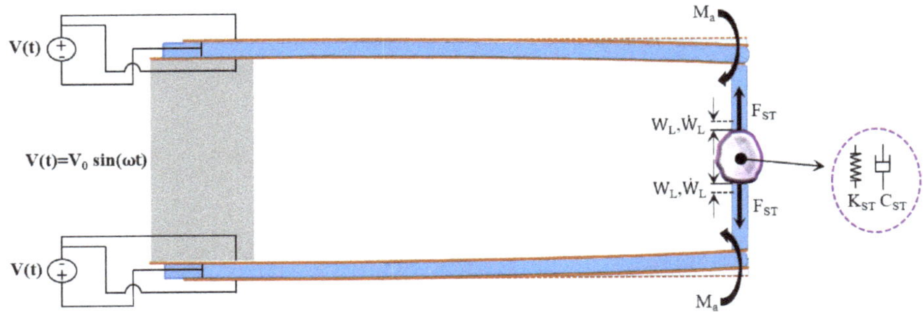

Figure 6. Scheme for the calculation of the stiffness and viscous damping of the soft tissue.

The viscous damping coefficient C_{ST} can be identified by measuring the amplitude $|w_L|$ in correspondence of the i-th vibration mode of the structure that has been excited by means of the piezoelectric elements, where the flexural modes of the beams are obtained as

$$\varphi(x) = B_1 \sin(\lambda x) + B_2 \cos(\lambda x) + B_3 \sinh(\lambda x) + B_4 \cosh(\lambda x) \tag{12}$$

with

$$
\begin{cases}
\varphi(0) = 0 \\[2mm]
\left. \dfrac{\partial \varphi(x)}{\partial x} \right|_{x=0} = 0 \\[2mm]
\left. \dfrac{\partial^2 \varphi(x)}{\partial x^2} \right|_{x=L} = 0 \\[2mm]
EI \left. \dfrac{\partial^3 \varphi(x)}{\partial x^3} \right|_{x=L} = K_{ST}\varphi(L) - \omega^2 m \varphi(L)
\end{cases}
\tag{13}
$$

By substituting Equation (12) in (13), a system of four equations in four unknowns (B_1, B_2, B_3, B_4) is obtained. The eigenfrequencies are obtained setting to zero the determinant of the matrix coefficient (with $\omega^2 = \lambda^4 \frac{EI}{\rho S}$) and then the eigenmodes can be calculated.

The i-th flexural mode can be excited by applying a potential function

$$V(t) = V_0 \sin(\omega_i t) \tag{14}$$

where $\varphi_i(x)$ and ω_i are the i-th flexural mode and its related frequency, respectively. In these conditions the displacement $w(x,t)$ and the virtual displacement $\tilde{w}(x,t)$ can be written as:

$$
\begin{cases}
w(x,t) = A_i(t)\varphi_i(x) \\[2mm]
\tilde{w}(x,t) = \varphi_i(x)
\end{cases}
\tag{15}
$$

By substituting Equations (15) and (10) in (9), the following equation is obtained:

$$
\begin{aligned}
E_b I_b A_i(t) \int_0^L \left(\frac{\partial^2 \varphi_i(x)}{\partial x^2} \right)^2 dx = -\rho S \, \ddot{A}_i(t) \int_0^L \varphi_i(x)^2 dx - m \, \ddot{A}_i(t) \, \varphi_i(L)^2 + \\
+ M_a \left. \frac{\partial \varphi_i(x)}{\partial x} \right|_{x=L} - K_{ST} A_i(t)\varphi_i(L)^2 - C_{ST} \, \dot{A}_i(t) \, \varphi_i(L)^2
\end{aligned}
\tag{16}
$$

assuming (see Equation (1)):

$$\begin{cases} M = \rho S \int_0^L \varphi_i(x)^2 dx + m\varphi_i(L)^2 \\[2mm] C = C_{ST}\varphi_i(L)^2 \\[2mm] K = E_b I_b \int_0^L \left(\frac{\partial^2 \varphi_i(x)}{\partial x^2}\right)^2 dx + K_{ST}\varphi_i(L)^2 \\[2mm] Q = \frac{\psi}{6+\psi} E_a c T_b d_{31} \left.\frac{\partial \varphi_i(x)}{\partial x}\right|_{x=L} V_0 \end{cases} \tag{17}$$

and therefore Equation (16) becomes:

$$M\ddot{A}_i(t) + C\,\dot{A}_i\,(t) + KA_i(t) = Q\sin(\omega_i t) \tag{18}$$

By neglecting the transient part (see Equation (15.1)), the amplitude of the free end displacement is:

$$\left|w_L\right| = A_{i_f}\varphi_i(L) \tag{19}$$

where

$$A_{i_f} = \frac{Q}{C\omega_i}. \tag{20}$$

Finally, the damping coefficient of the soft tissue

$$C_{ST} = \frac{Q}{\left|w_L\right|\omega_i\varphi_i(L)} \tag{21}$$

can be found from Equation (17.2).

4. Results and Discussion

To validate the proposed model, numerical simulations have been done. The dimensions and the material characteristics are summarized in Table 1:

Table 1. Beams, piezoelectric plates and end masses specifications; lengths are expressed in μm.

	Material	Length	Thickness	Width	Young's mod. [GPa]	Density [kg/m^3]
Beams	Silicon	750	2	80	170	2329
Piezoel. pl.	AlN	750	$2*10^{-2}$	80	350	3300
Clamp teeth	Silicon	36	7.5	80	170	2329

In this work, the FEM results have been chosen as the reference values. Three typical soft tissues (liver, muscle and uterus) of known characteristics [83], have been considered. The value of the beam tip displacement, obtained by the FEM simulations, has been included in the mathematical model to calculate the equivalent Young's modulus and the viscous damping coefficient. The ST sample stiffness can be calculated by means of Equation (8). The values of the Young's modulus and the viscous coefficients reported in the literature have been compared with the ones calculated by the new method. Because of the symmetry of the structure with respect to its mid-plane (see Figure 7) only the upper part has been considered in the simulations.

In Table 2 the results of the static simulations, obtained by the above described procedure, have been reported. It is possible to observe that the model results are in good agreement with the real values with a percentage error less than 6% in all the cases.

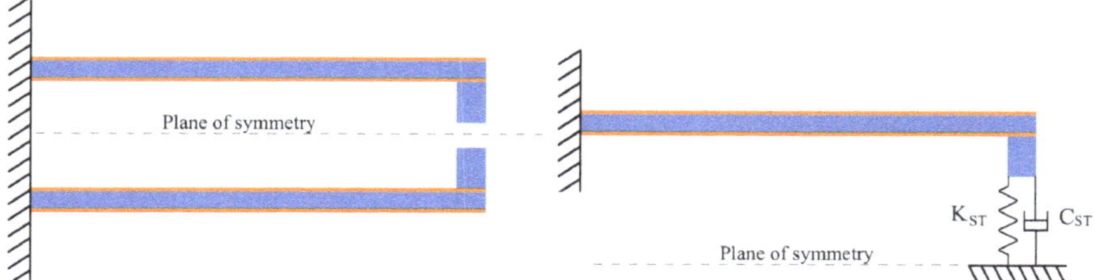

Figure 7. Scheme used for the simulations.

Table 2. Comparison between the actual Young's modules and those obtained by the model; E_{mm} is the Young's modulus obtained by the mathematical model.

	E [kPa]	E_{mm} [kPa]	*Err* %
Liver	10	9.47	5.3%
Muscle	20	19.43	2.8%
Uterus	30	29.64	1.2 %

To obtain the viscous damping coefficient, dynamics simulations are necessary. As described above, the chosen strategy consists in exciting, by the piezoelectric plates, the *i*-th mode of the structure, in order to obtain the amplitude of the free end displacement $|w_L|$ and then in including this in the model.

A comparison between the eigenfrequencies obtained by the COMSOL FEM code and the proposed model is reported in Table 3.

Table 3. Comparison between the eigenfrequencies obtained by the COMSOL FEM code and the proposed model.

Liver			
	FEM	*Model*	*Err%*
$f_1[Hz]$	15,523.50	15,856.40	2.14%
$f_2[Hz]$	30,701.40	29,487.60	3.95%
$f_3[Hz]$	82,116.70	75,668.50	7.85%
$f_4[Hz]$	162,877.10	151,620.00	6.91%

Muscle			
	FEM	*Model*	*Err%*
$f_1[Hz]$	18,768.20	18,686.20	0.44%
$f_2[Hz]$	35,437.60	34,681.70	2.13%
$f_3[Hz]$	82,674.80	76,297.90	7.71%
$f_4[Hz]$	162,995.20	151,781	6.88%

Uterus			
	FEM	*Model*	*Err%*
$f_1[Hz]$	20,066.80	19,721.70	1.72%
$f_2[Hz]$	40,190.2	39,660.70	1.32%
$f_3[Hz]$	83,255.70	77,038.00	7.47%
$f_4[Hz]$	163,118.00	151,949.00	6.85%

In the FEM simulations the first mode has been chosen to excite the structure with the values of the electrical potential reported in Table 4.

Table 4. Electrical potential used for the simulations.

	$V_0[V]$
Liver	8
Muscle	5
Uterus	5

By neglecting the initial transient part, the axial displacements for the various soft tissues have been reported in Figure 8.

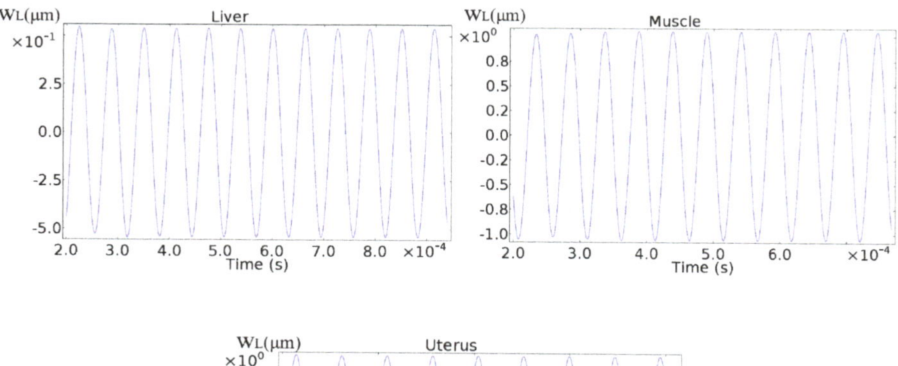
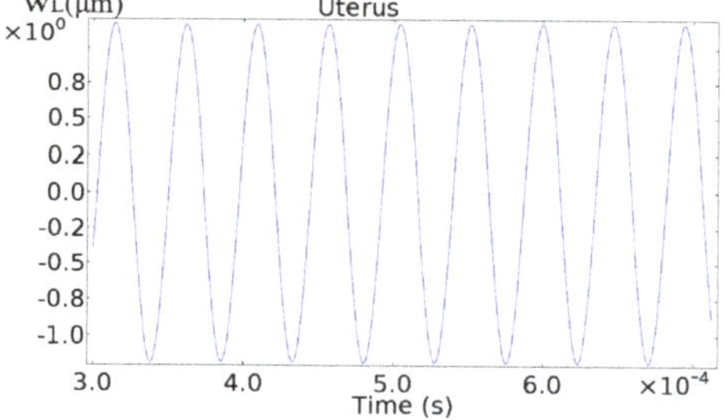

Figure 8. Axial displacement for the various soft tissues.

Finally, the comparison between the actual viscous damping coefficients and those obtained by the model results have been highlighted in Table 5.

Table 5. Comparison between the effective viscous damping and those obtained by the model.

	$C[Ns/m]$	$C_{mm}[Ns/m]$	$Err\%$
Liver	$3.07 * 10^{-5}$	$2.94 * 10^{-5}$	4.2%
Muscle	$5.09 * 10^{-5}$	$4.98 * 10^{-5}$	2.2%
Uterus	$7.14 * 10^{-5}$	$7.26 * 10^{-5}$	1.7%

A good agreement between the relative coefficients is also observed with a percentage error always less than 5%.

5. Conclusions

In this paper a novel piezo MEMS tweezer for soft materials characterization has been proposed. The tweezer mechanical structure is compatible with the known fabrication processes. A new mathematical model to calculate the stiffness, the equivalent Young's modulus and the viscous damping coefficient of the soft tissues is suggested. The method has been tested by comparing its results with Finite Element

Analysis based on experimental data from the literature. The two sets of data are in good agreement with a difference less than 6% in all the considered cases.

Author Contributions: F.B. has designed and developed the mathematical model. A.R. and F.B. carried out the numerical simulations. N.P.B. has reviewed the theoretical approach and the compatibility of the structure with some technological fabrication process.

Conflicts of Interest: The authors declare no conflict of interest.

Abbreviations

The following abbreviations are used in this manuscript:

C	damping coefficient
C_{ST}	damping coefficient of the soft tissue
d_{31}	piezoelectric coefficient
E_a	Young's modulus of the piezoelectric material
E_b	Young's modulus of the beam
E_{ST}	equivalent Young's modulus of the soft tissue
F_{ST}	force applied from the soft tissue to the tweezers
I_b	inertia moment of the beam
K	stiffness coefficient
K_{ST}	stiffness coefficient of the soft tissue
L	beam length
m	mass applied at the end of the tweezer
M_a	piezoelectric bending moment
M	mass coefficient
S_{ST}	cross-sectional area of the soft tissue
ST	soft tissue
T_a	piezoelectric thickness
T_b	beam thickness
V	electric potential applied to the piezoelectric plates
w	transverse displacement
w_L	transverse displacement of the free end
$\tilde{w}$	virtual transverse displacement
$\phi_i(x)$	i-th flexural mode of the beam
ρ	mass density of the beam
ω_i	natural frequency of the i-th flexural mode

References

1. Tang, M.; Xia, L.; Wei, D.; Yan, S.; Du, C.; Cui, H.L. Distinguishing Different Cancerous Human Cells by Raman Spectroscopy Based on Discriminant Analysis Methods. *Appl. Sci.* **2017**, *7*, 900. [CrossRef]
2. Rianna, C.; Radmacher, M. Comparison of viscoelastic properties of cancer and normal thyroid cells on different stiffness substrates. *Eur. Biophys. J.* **2017**, *46*, 309–324. [CrossRef] [PubMed]
3. Lekka, M.; Gil, D.; Pogoda, K.; Dulińska-Litewka, J.; Jach, R.; Gostek, J.; Klymenko, O.; Prauzner-Bechcicki, S.; Stachura, Z.; Wiltowska-Zuber, J.; et al. Cancer cell detection in tissue sections using AFM. *Arch. Biochem. Biophys.* **2012**, *518*, 151–156. [CrossRef] [PubMed]
4. Discher, D.E.; Boal, D.H.; Boey, S.K. Simulations of the erythrocyte cytoskeleton at large deformation. II. Micropipette aspiration. *Biophys. J.* **1998**, *75*, 1584–1597. [CrossRef]
5. Bausch, A.R.; Ziemann, F.; Boulbitch, A.A.; Jacobson, K.; Sackmann, E. Local measurements of viscoelastic parameters of adherent cell surfaces by magnetic bead microrheometry. *Biophys. J.* **1998**, *75*, 2038–2049. [CrossRef]
6. Weisenhorn, A.L.; Khorsandi, M.; Kasas, S.; Gotzos, V.; Butt, H.J. Deformation and height anomaly of soft surfaces studied with an AFM. *Nanotechnology* **1993**, *4*, 106. [CrossRef]
7. Guck, J.; Ananthakrishnan, R.; Cunningham, C.C.; Käs, J. Stretching biological cells with light. *J. Phys. Condens. Matter* **2002**, *14*, 4843. [CrossRef]

8. Adldoost, H.; Jouibary, B.R.; Zabihollah, A. Design of SMA micro-gripper for minimally invasive surgery. In Proceedings of the 2012 19th Iranian Conference of Biomedical Engineering, Tehran, Iran, 20–21 December 2012.

9. Botta, F.; Verotti, M.; Bagolini, A.; Bellutti, P.; Belfiore, N.P. Mechanical response of four-bar linkage microgrippers with bidirectional electrostatic actuation. *Actuators* **2018**, *7*, 78. [CrossRef]

10. Tsai, Y.C.; Lei, S.H.; Sudin, H. Design and analysis of planar compliant microgripper based on kinematic approach. *J. Micromech. Microeng.* **2004**, *15*, 143. . [CrossRef]

11. Wierzbicki, R.; Adda, C.; Hotzendorfer, H. Electrostatic silicon microgripper with low voltage of actuation. In Proceedings of the 2007 International Symposium on Micro-NanoMechatronics and Human Science, Nagoya, Japan, 11–14 November 2007.

12. Hamedi, M.; Salimi, P.; Vismeh, M. Simulation and experimental investigation of a novel electrostatic microgripper system. *Microelectron. Eng.* **2012**, *98*, 467–471. [CrossRef]

13. Piriyanont, B.; Moheimani, S.R.; Bazaei, A. Design and control of a MEMS micro-gripper with integrated electro-thermal force sensor. In Proceedings of the 2013 Australian Control Conference, Fremantle, WA, Australia, 4–5 November 2013.

14. Wierzbicki, R.; Houston, K.; Heerlein, H.; Barth, W.; Debski, T.; Eisinberg, A.; Dario, P. Design and fabrication of an electrostatically driven microgripper for blood vessel manipulation. *Microelectron. Eng.* **2006**, *83*, 1651–1654. [CrossRef]

15. Zhang, Y.; Chen, B.K.; Liu, X.; Sun, Y. Autonomous robotic pick-and-place of microobjects. *IEEE Trans. Robot.* **2010**, *26*, 200–207. [CrossRef]

16. Scalari, G.; Eisinberg, A.; Mazzoni, M.; Menciassi, A.; Dario, P. A sensorized μelectro discharge machined superelastic alloy microgripper for micromanipulation. In Proceedings of the IEEE/RSJ International Conference on Intelligent Robots and Systems, Lausanne, Switzerland, 30 September–4 October 2002; pp. 1591–1595.

17. Greminger, M.A.; Sezen, A.S.; Nelson, B.J. A four degree of freedom MEMS microgripper with novel bi-directional thermal actuators. In Proceedings of the 2005 IEEE/RSJ International Conference on Intelligent Robots and Systems, Edmonton, AB, Canada, 2–6 August 2005.

18. Chang, R.J.; Chen, C.C. Using microgripper for adhesive bonding in automatic microassembly system. In Proceedings of the International Conference on Mechatronics and Automation, Harbin, China, 5–8 August 2007.

19. Chronis, N.; Lee, L.P. Polymer MEMS-based microgripper for single cell manipulation. In Proceedings of the 17th IEEE International Conference on Micro Electro Mechanical Systems. Maastricht MEMS 2004 Technical Digest, Maastricht, The Netherlands, 25–29 January 2004.

20. Sun, X.; Chen, W.; Fatikow, S.; Tian, Y.; Zhou, R.; Zhang, J.; Mikczinski, M. A novel piezo-driven microgripper with a large jaw displacement. *Microsyst. Technol.* **2015**, *21*, 931–942. [CrossRef]

21. Belfiore, N.P.; Broggiato, G.B.; Verotti, M.; Balucani, M.; Crescenzi, R.; Bagolini, A.; Bellutti, P.; Boscardin, M. Simulation and construction of a mems CSFH based microgripper. *Int. J. Mech. Control* **2015**, *16*, 21–30.

22. Verotti, M.; Crescenzi, R.; Balucani, M.; Belfiore, N.P. MEMS-based conjugate surfaces flexure hinge. *J. Mech. Des.* **2015**, *137*, 012301. [CrossRef]

23. Cecchi, R.; Verotti, M.; Capata, R.; Dochshanov, A.; Broggiato, G.; Crescenzi, R.; Balucani, M.; Natali, S.; Razzano, G.; Lucchese, F.; et al. Development of micro-grippers for tissue and cell manipulation with direct morphological comparison. *Micromachines* **2015**, *6*, 1710–1728. [CrossRef]

24. Verotti, M.; Dochshanov, A.; Belfiore, N.P. Compliance Synthesis of CSFH MEMS-Based Microgrippers. *J. Mech. Des.* **2017**, *139*. [CrossRef]

25. Crescenzi, R.; Balucani, M.; Belfiore, N.P. Operational characterization of CSFH MEMS technology based hinges. *J. Micromech. Microeng.* **2018**, *28*, 055012. [CrossRef]

26. Verotti, M. Analysis of the center of rotation in primitive flexures: Uniform cantilever beams with constant curvature. *Mech. Mach. Theory* **2016**, *97*, 29–50. [CrossRef]

27. Verotti, M. Effect of initial curvature in uniform flexures on position accuracy. *Mech. Mach. Theory* **2018**, *119*, 106–118. [CrossRef]

28. Di Giamberardino, P.; Bagolini, A.; Bellutti, P.; Rudas, I.; Verotti, M.; Botta, F.; Belfiore, N.P. New MEMS tweezers for the viscoelastic characterization of soft materials at the microscale. *Micromachines* **2018**, *9*, 15. [CrossRef] [PubMed]

29. Ursi, P.; Santoro, A.; Gemini, A.; Arezzo, A.; Pironi, D.; Renzi, C.; Cirocchi, R.; Di Matteo, F.; Maturo, A.; D'Andrea, V.; et al. Comparison of outcomes following intersphincteric resection vs low anterior resection for low rectal cancer: A systematic review. *G. Di Chir.* **2018**, *39*, 123–142.

30. Balla, A.; Quaresima, S.; Ursi, P.; Seitaj, A.; Palmieri, L.; Badiali, D.; Paganini, A.M. Hiatoplasty with crura buttressing versus hiatoplasty alone during laparoscopic sleeve gastrectomy. *Gastroenterol. Res. Pract.* **2017**, *2017*. [CrossRef] [PubMed]

31. Cochetti, G.; Del Zingaro, M.; Boni, A.; Cocca, D.; Panciarola, M.; Tiezzi, A.; Gaudio, G.; Balzarini, F.; Ursi, P.; Mearini, E. Colovesical fistula: Review on conservative management, surgical techniques and minimally invasive approaches. *G. Di Chir.* **2018**, *39*, 195–207.

32. Popivanov, G.; Tabakov, M.; Mantese, G.; Cirocchi, R.; Piccinini, I.; D'Andrea, V.; Covarelli, P.; Boselli, C.; Barberini, F.; Tabola, R.; et al. Surgical treatment of gastrointestinal stromal tumors of the duodenum: A literature review. *Transl. Gastroenterol. Hepatol.* **2018**, *3*. [CrossRef] [PubMed]

33. Paci, M.; Scoglio, D.; Ursi, P.; Barchetti, L.; Fabiani, B.; Ascoli, G.; Lezoche, G. Transanal Endocopic Microsurgery (TEM) in advanced rectal cancer disease treatment [Il ruolo della TEM nel trattamento dei tumori del retto extraperitoneale]. *Ann. Ital. Di Chir.* **2010**, *81*, 269–274.

34. Quaresima, S.; Balla, A.; Dambrosio, G.; Bruzzone, P.; Ursi, P.; Lezoche, E.; Paganini, A.M. Endoluminal loco-regional resection by TEM after R1 endoscopic removal or recurrence of rectal tumors. *Minim. Invasive Ther. Allied Technol.* **2016**, *25*, 134–140. [CrossRef]

35. Lezoche, E.; Fabiani, B.; D'Ambrosio, G.; Ursi, P.; Balla, A.; Lezoche, G.; Monteleone, F.; Paganini, A.M. Nucleotide-guided mesorectal excision combined with endoluminal locoregional resection by transanal endoscopic microsurgery in the treatment of rectal tumors: Technique and preliminary results. *Surg. Endosc.* **2013**, *27*, 4136–4141. [CrossRef]

36. Legtenberg, R.; Groeneveld, A.W.; Elwenspoek, M. Comb-drive actuators for large displacements. *J. Micromech. Microeng.* **1996**, *6*, 320. [CrossRef]

37. Chen, T.; Sun, L.; Chen, L.; Rong, W.; Li, X. A hybrid-type electrostatically driven microgripper with an integrated vacuum tool. *Sens. Actuators A Phys.* **2010**, *158*, 320–327. [CrossRef]

38. Yeh, J.A.; Chen, C.N.; Lui, Y.S. Large rotation actuated by in-plane rotary comb-drives with serpentine spring suspension. *J. Micromech. Microeng.* **2004**, *15*, 201. [CrossRef]

39. Yeh, J.A.; Jiang, S.S.; Lee, C. MOEMS variable optical attenuators using rotary comb drive actuators. *IEEE Photonics Technol. Lett.* **2006**, *18*, 1170–1172. [CrossRef]

40. Kim, K.; Liu, X.; Zhang, Y.; Sun, Y. Nanonewton force-controlled manipulation of biological cells using a monolithic MEMS microgripper with two-axis force feedback. *J. Micromech. Microeng.* **2008**, *18*. [CrossRef]

41. Kim, K.; Liu, X.; Zhang, Y.; Cheng, J.; Wu, X.Y.; Sun, Y. Elastic and viscoelastic characterization of microcapsules for drug delivery using a force-feedback MEMS microgripper. *Biomed. Microdevices* **2009**, *11*, 421–427. [CrossRef]

42. Solano, B.; Wood, D. Design and testing of a polymeric microgripper for cell manipulation. *Microelectron. Eng.* **2007**, *84*, 1219–1222. [CrossRef]

43. Zeman, M.J.; Bordatchev, E.V.; Knopf, G.K. Design, kinematic modeling and performance testing of an electro-thermally driven microgripper for micromanipulation applications. *J. Micromech. Microeng.* **2006**, *16*, 1540. [CrossRef]

44. Zhang, R.; Chu, J.; Wang, H.; Chen, Z. A multipurpose electrothermal microgripper for biological micro-manipulation. *Microsyst. Technol.* **2013**, *19*, 89–97. [CrossRef]

45. Cauchi, M.; Grech, I.; Mallia, B.; Mollicone, P.; Sammut, N. The effects of cold arm width and metal deposition on the performance of a U-beam electrothermal MEMS microgripper for biomedical applications. *Micromachines* **2019**, *10*, 167. [CrossRef]

46. Cauchi, M.; Grech, I.; Mallia, B.; Mollicone, P.; Sammut, N. The effects of structure thickness, air gap thickness and silicon type on the performance of a horizontal electrothermal MEMS microgripper. *Actuators* **2018**, *7*, 38. [CrossRef]

47. Ren, H.; Wu, S.T. Adaptive Lenses Based on Soft Electroactive Materials. *Appl. Sci.* **2018**, *8*, 1085. [CrossRef]

48. Yan, B.; Wang, K.; Hu, Z.; Wu, C.; Zhang, X. Shunt damping vibration control technology: A review. *Appl. Sci.* **2017**, *7*, 494. [CrossRef]

49. Rossi, A.; Orsini, F.; Scorza, A.; Botta, F.; Belfiore, N.P.; Sciuto, S.A. A review on parametric dynamic models of magnetorheological dampers and their characterization methods. *Actuators* **2018**, *7*, 16. [CrossRef]

50. Botta, F.; Marx, N.; Gentili, S.; Schwingshackl, C.W.; Di Mare, L.; Cerri, G.; Dini, D. Optimal placement of piezoelectric plates for active vibration control of gas turbine blades: Experimental results. In Proceedings of the SPIE - The International Society for Optical Engineering, San Diego, CA, USA, 11–15 March 2012; Volume 8345.

51. Botta, F.; Dini, D.; Schwingshackl, C.; Di Mare, L.; Cerri, G. Optimal placement of piezoelectric plates to control multimode vibrations of a beam. *Adv. Acoustics Vib.* **2013**. . [CrossRef]

52. Botta, F.; Rossi, A.; Schinaia, L.; Scorza, A.; Orsini, F.; Sciuto, S.A.; Belfiore, N.P. Experimental validation on optimal placement of pzt plates for active beam multimode vibrations reduction. In Proceedings of the 23rd Conference of the Italian Association of Theoretical and Applied Mechanics AIMETA 2017, Salerno, Italy, 4–7 September 2017; Volume 3, pp. 2258–2269.

53. Botta, F.; Toccaceli, F. Piezoelectric plates distribution for active control of torsional vibrations. *Actuators* **2018**, *7*, 23. [CrossRef]

54. Xiao, Y.; Wang, B.; Zhou, S. Pull-in voltage analysis of electrostatically actuated MEMS with piezoelectric layers: A size-dependent model. *Mech. Res. Commun.* **2015**, *66*, 7–14. [CrossRef]

55. Azizi, S.; Ghodsi, A.; Jafari, H.; Ghazavi, M.R. A conceptual study on the dynamics of a piezoelectric MEMS (Micro Electro Mechanical System) energy harvester. *Energy* **2016**, *96*, 495–506. [CrossRef]

56. Madinei, H.; Khodaparast, H.H.; Adhikari, S.; Friswell, M.I.; Fazeli, M. Adaptive tuned piezoelectric MEMS vibration energy harvester using an electrostatic device. *Eur. Phys. J. Spec. Top.* **2015**, *224*, 2703–2717. [CrossRef]

57. Kanno, I. Piezoelectric MEMS for energy harvesting. *J. Phys. Conf. Ser.* **2015**, *660*. [CrossRef]

58. Tian, W.; Ling, Z.; Yu, W.; Shi, J. A review of MEMS scale piezoelectric energy harvester. *Appl. Sci.* **2018**, *8*, 645. [CrossRef]

59. Nah, S.K.; Zhong, Z.W. A microgripper using piezoelectric actuation for micro-object manipulation. *Sens. Actuators A Phys.* **2007**, *133*, 218–224. [CrossRef]

60. Haddab, Y.; Chaillet, N.; Bourjault, A. A microgripper using smart piezoelectric actuators. In Proceedings of the 2000 IEEE/RSJ International Conference on Intelligent Robots and Systems, Takamatsu, Japan, 31 October–5 November 2000; Volume 1.

61. Wang, D.H.; Yang, Q.; Dong, H.M. A monolithic compliant piezoelectric-driven microgripper: Design, modeling, and testing. *IEEE/ASME Trans. Mechatron.* **2013**, *18*, 138–147. [CrossRef]

62. Loh, O.; Vaziri, A.; Espinosa, H.D. The potential of MEMS for advancing experiments and modeling in cell mechanics. *Exp. Mech.* **2009**, *49*, 105–124. [CrossRef]

63. Crawley, E.F.; Luis, J.D. Use of piezoelectric actuators as elements of intelligent structures. *AIAA J.* **1987**, *25*, 1373–1385. [CrossRef]

64. Eom, C.B.; Trolier-McKinstry, S. Thin-film piezoelectric MEMS. *MRS Bull.* **2012**, *37*, 1007–1017. [CrossRef]

65. Xu, R.; Lei, A.; Christiansen, T.L.; Hansen, K.; Guizzetti, M.; Birkelund, K.; Thomsen, E.V.; Hansen, O. Screen printed PZT/PZT thick film bimorph MEMS cantilever device for vibration energy harvesting. In Proceedings of the 2011 16th International Solid-State Sensors, Actuators and Microsystems Conference, TRANSDUCERS'11, Beijing, China, 5–9 June 2011; pp. 679–682.

66. Shen, D.; Park, J.H.; Ajitsaria, J.; Choe, S.Y.; Wikle, H.C.; Kim, D.J. The design, fabrication and evaluation of a MEMS PZT cantilever with an integrated Si proof mass for vibration energy harvesting. *J. Micromech. Microeng.* **2008**, *18*, 055017. [CrossRef]

67. Mayrhofer, P.M.; Wistrela, E.; Kucera, M.; Bittner, A.; Schmid, U. Fabrication and characterisation of ScAlN-based piezoelectric MEMS cantilevers. In Proceedings of the 2015 Transducers—2015 18th International Conference on Solid-State Sensors, Actuators and Microsystems, TRANSDUCERS 2015, Anchorage, AK, USA, 21–25 June 2015; pp. 2144–2147.

68. Nagano, T.; Nishigaki, M.; Abe, K.; Itaya, K.; Kawakubo, T. Fabrication and performance of piezoelectric MEMS tunable capacitors constructed with AlN bimorph structure. In Proceedings of the IEEE MTT-S International Microwave Symposium Digest, San Francisco, CA, USA, 11–16 June 2006; pp. 1285–1288.

69. Park, J.S.; Yang, S.J.; Kyung-Il, L.E.E.; Kang, S.G. Fabrication and electro-mechanical characteristics of piezoelectric micro bending actuators on silicon substrates. *J. Ceram. Soc. Jpn.* **2006**, *114*, 1089–1092. [CrossRef]

70. Yuan, Y.H.; Du, H.J.; Xia, X.; Wong, Y.R. Modeling, fabrication and characterization of piezoelectric ZnO-based micro-sensors and micro-actuators. *Appl. Mech. Mater.* **2014**, *444–445*, 1636–1643. [CrossRef]

71. Lou, L.; Yu, H.; Haw, M.T.X.; Zhang, S.; Gu, Y.A. Comparative characterization of bimorph and unimorph AlN piezoelectric micro-machined ultrasonic transducers. In Proceedings of the IEEE International Conference on Micro Electro Mechanical Systems (MEMS), Shanghai, China, 24–28 January 2016; Volume 2016-February, pp. 1090–1093.

72. Zeng, Y.; Groenesteijn, J.; Alveringh, D.; Steenwelle, R.J.A.; Ma, K.; Wiegerink, R.J.; Lotters, J.C. Micro Coriolis MASS flow sensor driven by integrated PZT thin film actuators. In Proceedings of the IEEE International Conference on Micro Electro Mechanical Systems (MEMS), Belfast, UK, 21–25 January 2018; Volume 2018-January, pp. 850–853.

73. Sindhanaiselvi, D.; Shanmuganantham, T. Investigation on performance of piezoelectric beam based MEMS actuator for focussing of micro lens in mobile application. In Proceedings of the IEEE International Conference on Circuits and Systems, ICCS 2017, Thiruvananthapuram, India, 20–21 December 2017; Volume 2018-January, pp. 398–403.

74. Chen, S.H.; Michael, A.; Kwok, C.Y. A Fast Response MEMS Piezoelectric Microlens Actuator with Large Stroke and Low Driving Voltage. In Proceedings of the NEMS 2018 - 3th Annual IEEE International Conference on Nano/Micro Engineered and Molecular Systems, Singapore, Singapore, 22–26 April 2018; pp. 199–203.

75. Tsaur, J.; Zhang, L.; Maeda, R.; Matsumoto, S.; Khumpuang, S.; Wan, J. Design and fabrication of 1D and 2D micro scanners actuated by double layered PZT bimorph beams. In Proceedings of the 2001 International Microprocesses and Nanotechnology Conference, MNC 2001, Shimane, Japan, 31 October–2 November 2001; pp. 204–205.

76. Tsaur, J.; Zhang, L.; Maeda, R.; Matsumoto, S.; Khumpuang, S. Design and Fabrication of 1D and 2D Micro Scanners Actuated by Double Layered Lead Zirconate Titanate (PZT) Bimorph Beams. *Jpn. J. Appl. Phys.* **2002**, *41*, 4321–4326. [CrossRef]

77. Kuo, C.L.; Lin, S.C.; Wu, W.J. Fabrication and performance evaluation of a metal-based bimorph piezoelectric MEMS generator for vibration energy harvesting. *Smart Mater. Struct.* **2016**, *25*, 1280–1284. [CrossRef]

78. Che, L.; Halvorsen, E.; Chen, X.; Yan, X. A micromachined piezoelectric PZT-based cantilever in d33 mode. In Proceedings of the 2010 IEEE 5th International Conference on Nano/Micro Engineered and Molecular Systems, NEMS 2010, Xiamen, China, 20–23 January 2010; pp. 785–788.

79. Kanno, I. Piezoelectric MEMS: Ferroelectric thin films for MEMS applications. *Jpn. J. Appl. Phys.* **2018**, *57*. [CrossRef]

80. Strambi, G.; Barboni, R.; Gaudenzi, P. Pin-force and Euler-Bernoulli models for analysis of intelligent structures. *AIAA J.* **1995**, *33*, 1746–1749. [CrossRef]

81. Waisman, H.; Abramovich, H. Active stiffening of laminated composite beams using piezoelectric actuators. *Compos. Struct.* **2002**, *58*, 109–120. [CrossRef]

82. Botta, F.; Scorza, A.; Rossi, A. Optimal piezoelectric potential distribution for controlling multimode vibrations. *Appl. Sci.* **2018**, *8*, 551. [CrossRef]

83. Wells, P.N.; Liang, H.D. Medical ultrasound: Imaging of soft tissue strain and elasticity. *J. R. Soc. Interface* **2011**, *8*, 1521–1549. [CrossRef]

Article

Toward Operations in a Surgical Scenario: Characterization of a Microgripper via Light Microscopy Approach

Federica Vurchio [1], Pietro Ursi [2], Francesco Orsini [1], Andrea Scorza [1], Rocco Crescenzi [3], Salvatore A. Sciuto [1] and Nicola P. Belfiore [1,*]

[1] Department of Engineering, University of Roma Tre, via della Vasca Navale 79, 00146 Rome, Italy; federica.vurchio@uniroma3.it (F.V.); francesco.orsini@uniroma3.it (F.O.); andrea.scorza@uniroma3.it (A.S.); salvatore.sciuto@uniroma3.it (S.A.S.)
[2] Department of General Surgery and Surgical Specialties "Paride Stefanini", Sapienza University of Rome, Viale del Policlinico 155, 00161 Rome, Italy; pietro.ursi@uniroma1.it
[3] Department of Information Engineering, Electronic and Telecommunications, Sapienza University of Rome, Via Eudossiana 18, 00184 Roma, Italy; rocco.crescenzi@uniroma1.it
* Correspondence: nicolapio.belfiore@uniroma3.it; Tel.: +39-065733-3316

Received: 24 March 2019; Accepted: 2 May 2019; Published: 9 May 2019

Abstract: Micro Electro Mechanical Systems (MEMS)-Technology based micro mechanisms usually operate within a protected or encapsulated space and, before that, they are fabricated and analyzed within one Scanning Electron Microscope (SEM) vacuum specimen chamber. However, a surgical scenario is much more aggressive and requires several higher abilities in the microsystem, such as the capability of operating within a liquid or wet environment, accuracy, reliability and sophisticated packaging. Unfortunately, testing and characterizing MEMS experimentally without fundamental support of a SEM is rather challenging. This paper shows that in spite of large difficulties due to well-known physical limits, the optical microscope is still able to play an important role in MEMS characterization at room conditions. This outcome is supported by the statistical analysis of two series of measurements, obtained by a light trinocular microscope and a profilometer, respectively.

Keywords: microactuators; microgrippers; MEMS; displacement measurement; comb-drives; microscopy; profilometer; characterization; minimally invasive surgery

---

1. Introduction

The recent developments of the microsystems have been so promising that nowadays they offer a great potential to many applications which require a high grade of miniaturization. Nevertheless, using microsurgery to heal diseases with a minimal invasive approach is still an ambitious challenge because of severe requirements (accuracy, precision, reliability, reduced consumption, limited costs, small size, high performance repeatability, short response time, efficiency in wet or liquid environments). One way of coping with this challenge consists in modifying the common Micro Electro Mechanical Systems (MEMS) to increase their degrees of freedom (usually unitary) and dexterity, for example, providing them with several revolute pairs. These microsystems need to be tested in a significant environment, starting from room environment (in air), but this is not so obvious as it could appear at first sight. In fact, these systems need to be analyzed by means of SEM, which implies putting the mechanisms within a vacuum chamber, rather than a particular environment. Therefore, significant operational tests must be performed by other means of observations, such as microscopy or interferometric profilometry. However, the latter instruments have less resolution than a SEM and their use can be critical to inspect the smallest parts of a microsystem.

This paper shows that microscopy observation can still be very effective in MEMS operational tests under environmental conditions. An effort has been made to assess microscopy observation of MEMS by comparing this means with a higher class, but also a more expensive and difficult-to-use instrument, namely, a profilometer. The statistical treatment of the results of two experimental campaigns (optical microscope and profilometer) shows that the characterization and image acquisition capability of the optical microscope is comparable to that of the profilometer, while the optical microscope maintains a larger degree of freedom in setting the operational parameters.

Given the interdisciplinary nature of the present investigation, it is helpful to give a glimpse to the different types of involved competences. It is particularly important to review some previous contributions concerning the operational conditions that would be required to a micro-electro-mechanical system in surgery.

A first survey of MEMS for surgical applications [1] showed how MEMS technology may improve the functionality of existing surgical devices and also add new capabilities that give rise to new treatments. For example, MEMS can improve surgical outcomes, with lower risk, by providing the surgeon with real-time feedback on the operation. From the mechanical and operational point of view, microgrippers for different applications have been extensively proposed in literature [2,3].

There are many other different applications where MEMS can play a significant role. For a representative example, MEMS-Technology based micro-accelerometers [4] are able to measure the heart wall motion of patients who have undergone coronary artery bypass graft surgery (CABG). Furthermore, there are many other applications where MEMS could be conveniently used to improve success in surgery, such as in laparoscopic sleeve gastrectomy (LSG) with cruroplasty [5], in surgical treatment of gastrointestinal stromal tumors of the duodenum [6] in colovesical fistula surgery with minimally invasive approach [7], in Endoluminal loco-regional resection by Transanal Endoscopic Microsurgery (TEM) [8–10], and in Low Rectal Anterior Resection (LAR) [11].

These kinds of applications require very specific actions during the different stages of design, fabrication and packaging, because their correct working is depending on many physical parameters, such as temperature, humidity of the environment, presence of water or other liquids, chemical reactions and pressure. For example, the influence of temperature and humidity on the adsorbed water layer on micro scale monocrystalline silicon (Si) films has been investigated in air, using an Si-MEMS kHz-frequency resonator [12]. Water proof or water insensitive three-axis MEMS based accelerometers have been presented [13] for encouraging their operation in laboratory scale experiments. The theory of water electrolysis in a closed electrochemical cell has been described [14] to develop a new actuation principle for MEMS. A special apparatus has been employed to study the adhesive friction due to water in the nanometer range, where the water layer thickness greatly affects friction and adhesion [15]. Heat transfer characteristics of isolated bubble of water were investigated by local wall temperature measurement using a novel MEMS based sensor [16]. MEMS-OR-PAM (optical-resolution photoacoustic microscopy) has also been proved [17] to be a powerful tool for studying highly dynamic and time-sensitive biological phenomena.

Although MEMS still have not developed their potential for surgery applications, there are many more examples of their use in biomedical instruments [18–20]. The mechanical characteristics of cells provide information on their functionality and their state of health, while their geometry can be linked to genetic alterations and apoptosis; the importance of these analyses could help in the collection of phenotypic information or for the diagnosis of pathological disorders [21]. For example, a study[22] was conducted about the viscoelasticity of L929 cells, to investigate their cellular structure, notoriously linked to important physiological functions. In addition, applications of microspectroscopic techniques [23] have been conducted to characterize the viscoelastic properties of living cells. The study of the mechanical properties of cells, also concerns the muscle cells: for example the mechanical response, in terms of strength and displacement, of some smooth muscle cells subjected to elongation has been studied [24] with the aim of better understanding the links between their functionalities and their structure. A study [25] on skeletal muscle cells described how the relationship between the stiffness and the force exerted by some actin filaments can indicate the physiological state of the actin cytoskeleton. Moreover, the mechanical properties of tissue-engineered vascular constructs have been studied by monitoring pressure and diameter variations of vascular constructs submitted to hydrostatic loading [26]. A silicon microgripper has also been proposed [27] to characterize the mechanical stiffness of biological tissues, with the wider prospective of developing a professional inspecting instrument for laboratory measurements or surgical operations.

More recently, new emerging MEMS-Technology based instruments for biomedical and surgical applications have been developed. In fact, based on a new concept hinge [28,29], a class of different microgrippers, equipped with rotary comb-drives, has been developed [30–33], also in significant environments [34], and fabricated [35,36]. Thanks to their size, these instruments are expected to be employed in surgery and diagnostics. This class of microsystem is the focus of the present investigation.

2. Experimental Characterization of MEMS-Technology Based Instruments in Operational Environments

In spite of MEMS' great potential for biomedical applications, only a few studies [37] focus on their characterization in an operational environment.

In fact, SEM observation is often a necessity to characterize MEMS, but this means it could be detrimental for the assessment of tests under real operational conditions, basically because tests are performed in vacuum conditions.

Considering the extensively spread biomedical applications, it is also worth mentioning the Optical Coherence Tomography (OCT), which is an optical imaging technique used primarily to investigate the internal microstructure of biological tissues, in ophthalmology and in dentistry [38]. The main advantage of OCT compared to other traditional systems is that this method offers the possibility of a non-invasive in vivo visualization of the tissue with high-resolution three-dimensional images [39]. Despite the axial resolution of some OCT systems (such as Ultrahigh-resolution OCT) [40] that can reach a few micrometers [41], the lateral resolution is affected by the diffraction limits due to the spot size in the focal plane of the probe beam. In fact, the lateral resolution often presents values between 10 and 20 µm [42,43]. Indeed, the above mentioned values of lateral resolution, together with higher costs, make OCTs less attractive than the light microscopy for the characterization of microgrippers in operational conditions.

Considering the above mentioned characteristics for SEM and OCT, traditional light microscopy is quite competitive since it allows test stands to be less expensive in a widespread range of test conditions.

In this investigation a performance characterization of some comb-driven microgrippers in operative condition has been carried out.

The characterization of MEMS-Technology based microgrippers has been recently approached by using a profilometer [44], while the present work will show that a more simple commercial light microscope is still capable to satisfy the same activity with no detriment of measurement significance. Since optical microscopy is widely used in laboratories, it is believed that this paper might have some impact on the future characterizations of microsystems in air, wet or liquid environments. The importance of micromanipulation of mobile micro-particle suspended in liquid well has been recently underlined and a visual-servo automatic micromanipulating system has been presented [45].

Since the intrinsic resolution of optical microscopy is generally rather worse than SEM or profilometer imaging, its real aptitude and efficacy in examining micro-objects remains something that must be validated. Therefore, the present paper will endorse microscopy observation to characterize a microsystem in environmental conditions, by comparing the results obtained by means of both optical microscope and profilometer image analysis techniques. This is justified since the optical devices are the most suitable for carrying out measurements on microsystems without contact and in operative conditions.

In this work, the above mentioned characterization involved the measurements of the angular displacements of a rotary comb-drive embedded into an independently developed microgripper, when a voltage is applied to the device. This is obtained by an in-house Image Analysis Software (IAS) implemented by the authors. The analysis of the measurement results has been achieved by means of uncertainty models for the evaluation of the corresponding error sources. For this purpose, the two image acquisition systems and the corresponding measurement chains will be described, together with the analysis of the relative sources of uncertainty. Finally, the measurements carried out with the Optical Profilometer System (OPS) and Light Trinocular Microscope (LTM) systems will be analyzed to check whether the two groups of results could be considered comparable and consistent within the interval of experimental uncertainties.

3. Materials and Methods

The evolution of MEMS gave rise to different kinds of microsystems that have been based on MEMS Technology. However, despite the relevant elements mentioned above, one cannot help but notice that, in the international technical-scientific panorama, guidelines, protocols and regulations regarding the characterization of these devices are still lacking, both under the metrological and the mechanical point of view. The present investigation is part of a larger project dedicated to the development of new concept microgrippers for surgical applications and it attempts to partially fill this gap and to validate the optical image analysis as a proper means of characterization of the developed microsystems under operative conditions.

The peculiar object of this investigation is a rotary comb-drive depicted in Figure 1. This component consists in an electrostatic actuator that provides motion to the microgripper illustrated in Figure 1a. The rotary comb-drive is composed of a pair of sets of fingers, as shown in the more detailed Figure 1b. The mobile series of fingers rotates as a function of the applied voltage and the determination of the voltage-rotation curve is greatly important for the operational aspects. Two measurement chains have been set up to measure such voltage-rotation function. The first is composed by a Fogale Zoomsurf 3D Optical Profiling System (OPS, Table 1), while the second is composed of a NB50TS Eurotek Light Trinocular Microscope (LTM, Table 2).

For the sake of the present investigation it is convenient to underline that the two sets of images from the two devices have been both processed using the same in-house built software in order to properly discriminate which one, between OPS and LTM, is the most suitable system to characterize the microgripper devices.

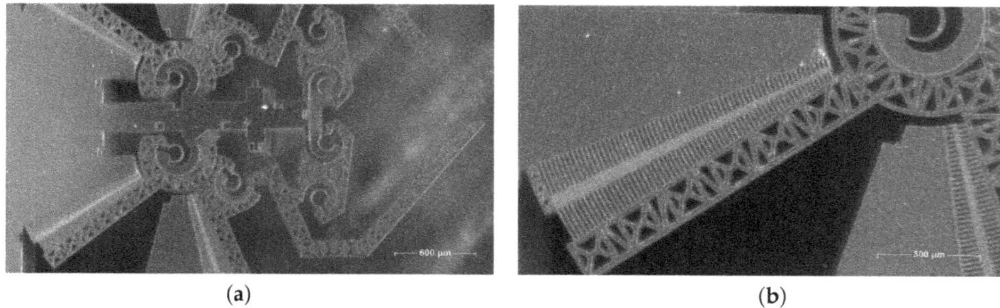

(a)   (b)

Figure 1. Image of a microgripper prototype obtained from optical microscope (**a**) and a detail of the rotary comb-drive (**b**).

Table 1. Optical Profilometer System (OPS) experimental setup: specifications of the comb-drive actuated silicon microgripper (Device Under Test, DUT) and of the testing stand.

Device	Characteristics
DUT Material	Silicon type P, dopant Boron, orientation <100>, electrical resistivity 0.005–0.030 Ohm·cm
DUT Geometry	module dimensions 2000 µm ×1500 µm, device thickness 40 µm, insulated layer thickness 3 µm, handle thickness 400 µm
Power Supply	Keithley 236, Range settable to 1.1/11/110/1100 V with respectively 0.1/1/10/100 mV resolution, accuracy 0.06 V (Range ±110.00 V)
Micropositioner	n.1 MP25L, n.1 MP25R, range X/Y/Z 10/10/10 mm with 5 µm resolution
Probes (supply)	PA-C-1M with tungsten needles
DUT Stage	The wafer containing the DUT is placed on the profilometer working surface and fixed by an adhesive tape
3D Optical Profilometer	Fogale Zoomsurf 3D optical profiling system, field of view from 7.2 mm × 5.4 mm to 80 nm × 60 nm, maximum lateral resolution 0.6 µm
Digital Image	768 × 580 pixels, 8 bit, 0.6 px/µm
Image Processing Software	In-house software developed in MATLAB (2017a, MathWorks)
Notebook pc	Intel core i7-2670, 6 Gb RAM, Nvidia GeForce GT 520 MX

Table 2. Light Trinocular Microscope (LTM) experimental setup (DUT, micropositioner, probes and PC as in Table 1).

Device	Characteristics
Power Supply	HP E3631A, DC Output: 0 to +25 V, 0 to −25 V, Resolution 1.5 mV, Accuracy 0.04 V at F.S.
DUT Stage	Instrumented support with micrometric screws for angular and linear movement of the sample, in the 3 orthogonal directions in space (x, y, z)
Light microscope	Eurotek NB50TS NB SOTS, Zoom range 0.8× …5× (8× …50×), LED illumination Transmitted-Reflected, B2-1525 additional objective 2×
Digital Image	1920 × 1080 pixels, 24 bit, 1.359 px/µm
Digital camera	MD6iS, 6MP, pixel dimension, 2.8 µm × 2.8 µm, maximum resolution 3264 × 1840 px
Image Processing Software	In-house software developed in MATLAB environment (2017a, MathWorks)

3.1. OPS and LTM Experimental Setups

The Device Under Test (DUT) consists of a microgripper prototype made up of pseudo-rigid beams and flexure hinges. Any jaw of the microgripper is actuated by a capacitive rotary comb-drive that provides a torque when a voltage is applied to the electrodes.

During the first experimental campaign, performed with the Optical Profiling System, a variable voltage source has been used to supply the DUT and to evaluate the angular displacements of its comb drive. In this investigation, a power supply Keithley 236 SMU with 0.06 V output voltage accuracy (Range ±110.00 V) has been used (Figure 2).

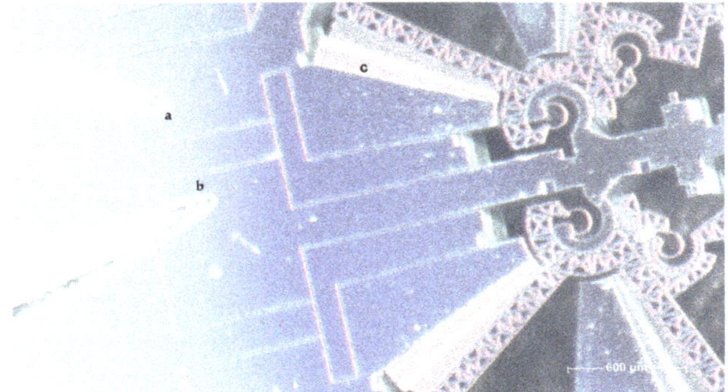

Figure 2. The two micropositioners with probe arms and tungsten needles (*a*) and (*b*), used in both the experimental setups; a voltage is applied to the comb-drive (*c*) of the DUT, in order to supply the device.

Two micropositioners with 5 μm resolution have been used, each one being equipped with tungsten needles, in order to apply the voltage to the DUT electrodes, Figure 3, and a set of digital images have been acquired using a Fogale Zoomsurf 3D OPS. The digital image resolution of 0.6 pixel/μm is provided by the Profilometer embedded software.

Figure 3. The two micropositioners with probe arms and tungsten needles.

During the image acquisition campaign performed by means of the LTM system the same microgripper sample has been analyzed. The power supply device is a HP E3631A, with 0.04 V output voltage accuracy (Range 0 to +25 V). An electric protection circuit, equipped with a fuse, has also been connected in series, between the power supply and the DUT, to cautiously prevent the passage of a current exceeding 200 mA, which could compromise the device. Two micropositioners have been used (Figure 3) and the microgripper angular displacements have been measured by means of acquired images and collected by a NB50TS Eurotek trinocular microscope system. The optical resolution is limited by the diffraction of the visible light and it is about 0.45–0.6 μm, as in conventional light microscopes [46]; in this study, the worst case of 0.6 μm has been considered. The pixel resolution has been provided by means of calibration procedure; by means of Matlab software, a length of 1000 μm on an image of a micrometer slide was considered and 15 tests were performed. By means of this procedure, the pixel density of 1.359 ± 0.007 pixel/μm was calculated. Furthermore, the sampling constraints (Nyquist limit) and the density of the photo sites on the digital sensor also limit the whole systems resolution. To achieve a *higher level* particle characterization (i.e., differentiation based upon higher order measurements such as circularity), the size of the particle or the object under examination must be greater than 4 μm [46]. In our case, in fact, it is possible to verify that it is impossible to resolve the comb-drive finger gap, that is 3 μm. However, the distance between two fingers, that is 10 μm, is clearly discriminated (see also Ref. [27]); for all these reasons, a overall resolution of about 4 μm for the LTM system is assumed. In Figure 4, the entire LTM setup is shown and some specifications of its main components are reported in Table 2. Once the device has been positioned on the DUT stage, it has been powered by means of two probes with tungsten needles. Through the micropositioners, the tungsten needles are approached to the microgripper electrical connections.

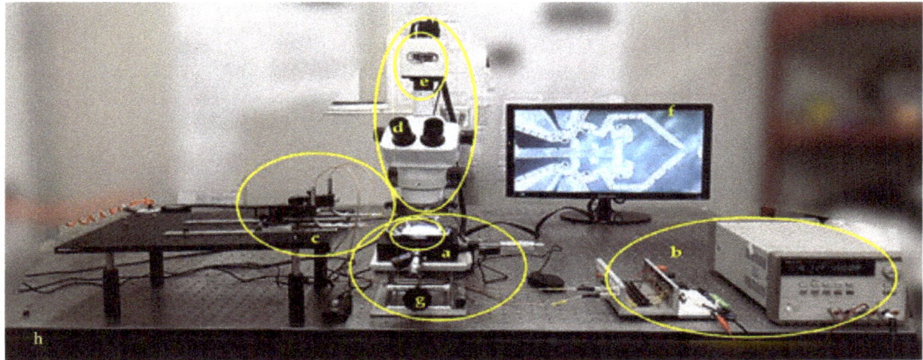

Figure 4. Optical Microscope measurement setup. Microgripper prototype (*a*), Supply voltage with protection circuit (*b*), micropositioners with two embedded probe arms and tungsten needles (*c*), Optical Microscope (*d*), embedded camera for images acquisition (*e*), a monitor for displaying and monitoring the device movement in operating conditions (*f*), instrumented support with micrometric screws (*g*), pneumatic suspension table (*h*).

Considering the OPS measurement chain, a set of images has been acquired, each one corresponding to a specific voltage setting: to calculate the angular displacement of the comb-drive, the first acquired image referred to 0 V, has been compared with the others (i.e., 2 V, 4 V, ..., 28 V); Figure 5a illustrates an example of an image that has been processed by the in-house built software. Considering the LTM measurement chain, the same steps have been carried out. However, in order to calculate the angular displacements, the first acquired image (0 V) has been compared with the other images referred to voltages up to 24 volts (i.e., 2 V, 4 V, ..., 24 V) instead of 28 V (as in the previous case); Figure 5b shows an example of image that has been processed by the in-house software. Considering the OPS, a set of 12 images has been acquired at each voltage, for a total of 180 images; considering the LTM image acquisition system, a set of 16 images has been acquired at each voltage, for a total of 208 images.

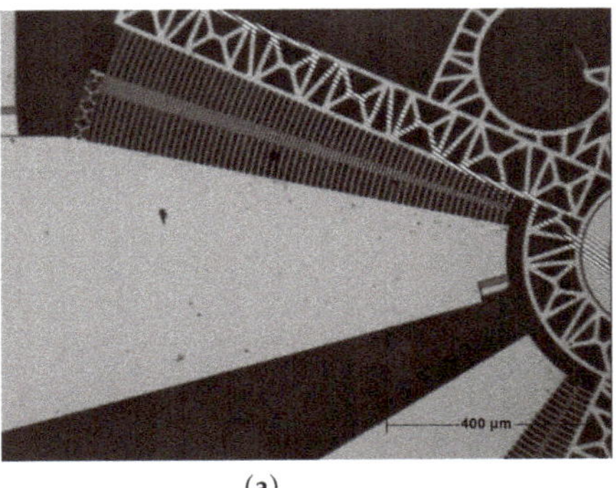
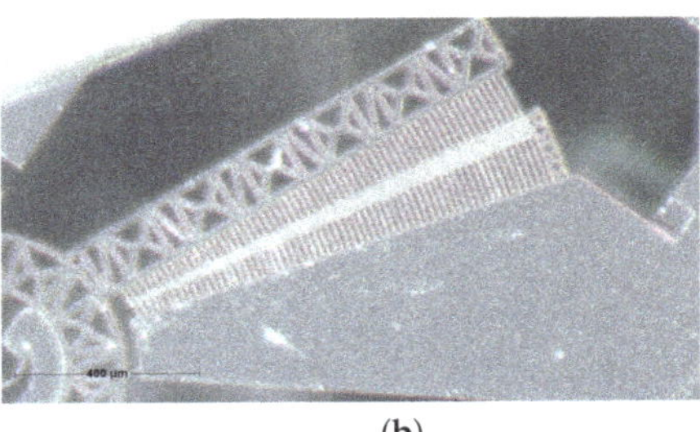

(a)　　　　　(b)

Figure 5. Two examples of images of the Microgripper comb-drive, acquired by optical profilometer (**a**), and optical microscope (**b**); both images (**a**,**b**) are referred to 0 V.

The OPS experimental investigation was made before the LTM campaign. Therefore, the testing voltages used in the second campaign have been influenced by the need of not imposing values that could damage the comb-drives, avoiding the pull-in effect. Such phenomenon occurred during the OPS campaign when the applied voltage was greater than 25 V. Above this value the fingers were unstable; in fact, they were used to get in contact quite often, inducing a short circuit.

Despite that the LTM system has a higher digital resolution of the acquired images, its Signal-to-Noise Ratio (SNR) is lower than the OPS SNR. This characteristic can be related to multiple factors, such as environmental noise, light source (i.e., non-homogeneous light source), optical aberrations, image A/D conversion and processing.

3.2. Image Analysis Software

A semi-automatic software has been implemented in Matlab according to the following steps.

1. At first, it is important to find the Instant Center of Rotation (ICR) of the DUT comb-drive. Considering the image that corresponds to 0 V (both for OPS and for LTM systems), four point are manually selected (a, a', b, b' and c, c', d, d') on the edges of the comb-drive and the ICR has been found as the intersection of the two lines (Figure 6).

2. The second step regards the manual selection of a particular Region Of Interest (ROI) on the same image considered previously. The identified ROI is characterized by a particular pixel pattern, which is compared with subsequent images (that correspond to 2 V, 4 V, ..., 28 V for OPS and 2 V, 4 V, ..., 24 V for LTM) to identify where this pattern is located. This procedure is carried out through a template-matching algorithm, which identifies for each image the coordinates of the point corresponding to the center of gravity (COG) of the ROI. This point corresponds to one of the three vertices of the triangle necessary to identify the angular aperture of the device comb-drive. The comb-drive has a static part fixed to the structure of the MEMS device and a mobile part. The ROI has been chosen on the mobile part of the microgripper comb-drive (Figure 7).

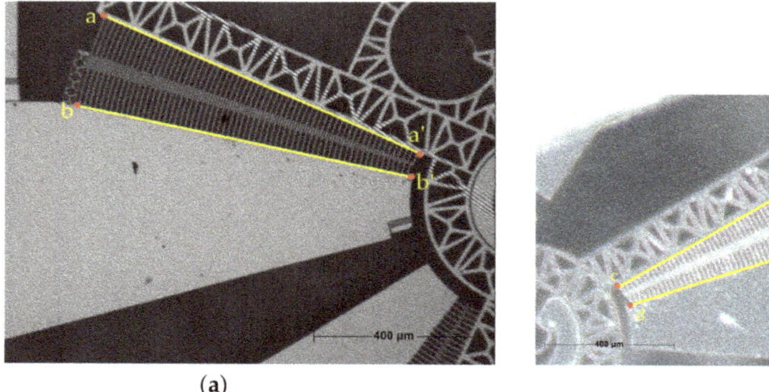

(a)   (b)

Figure 6. Two images of the microgripper comb-drive acquired by optical profilometer (**a**) and microscope (**b**), respectively.

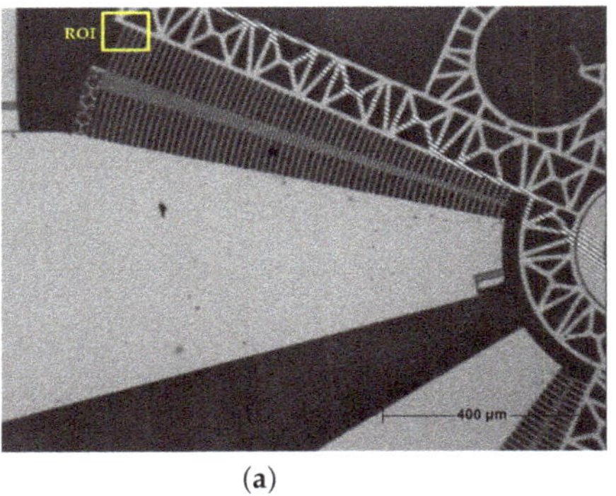
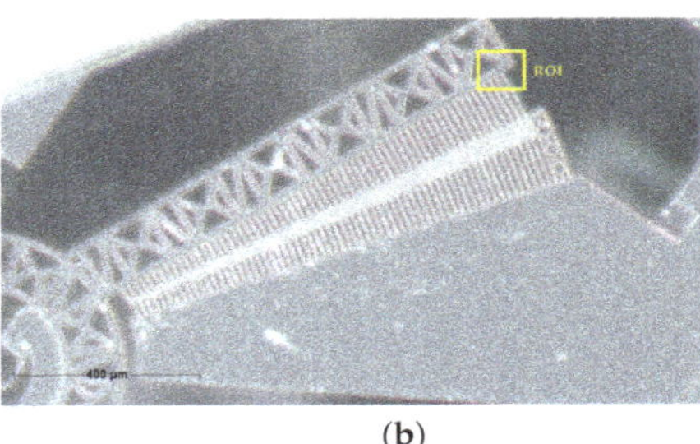

(a)   (b)

Figure 7. Manual selection of a particular region of the image (ROI), to find its center of gravity (COG) on OPS image (**a**) and on LTM image (**b**).

The first and second above described steps of the implemented software are the only ones in which the manual selection of an operator is necessary to properly select points on the images and the positions and dimensions of the ROI. The following ones have been automated.

3. As shown in Figure 8, the automatic part of the implemented software considers the coordinates of the following three points on the images:

   - The most distant point from the ICR on the fixed part of the comb-drive (a);
   - The ICR of the comb-drive (b);
   - The center of the ROI (c);

   The first two points remain fixed for all the following images, under the hypothesis of no deformation due to the movement of the comb-drive, while the only point, whose coordinates change for each considered image, is the center of the ROI.

4. These three points determine a triangle, where the vertex ICR corresponds to the angular opening α of the comb-drive. With reference to Figure 9, points a, b and c are considered to be the vertices of an isosceles triangle, where

$$\alpha = 2\arcsin\left(\frac{A}{B}\right) \tag{1}$$

   with $A = \frac{\overline{ca}}{2}$ and $B = \overline{cb}$. Using a template matching algorithm, a match is found between the coordinates of the center of the selected ROI on the first image (i.e., that corresponds to 0 V supply) and on all the subsequent images. Through this operation the in-house software detects the new coordinates of the ROI center of gravity (ROIGC) for each subsequent image and therefore for each applied voltage. The ROIGC changes its coordinates and, consequently, it changes also the shape of the corresponding triangle and the angular aperture of the DUT comb-drive, depending on the different voltage supply.

5. The angular displacement is obtained from the comb-drive angular aperture at each voltage value.

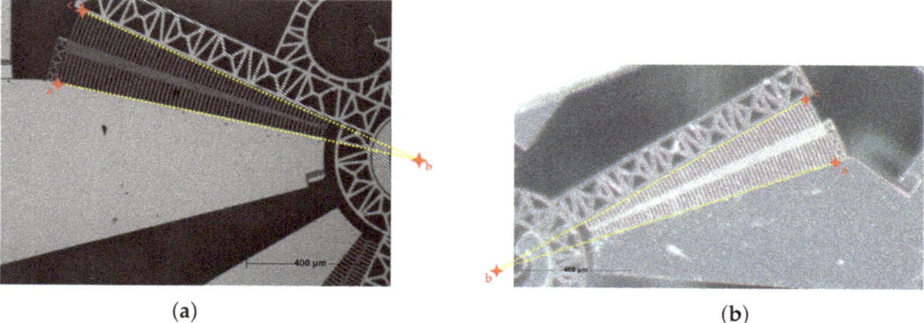

(a) (b)

Figure 8. The determination of three points on the profilometer (**a**) and the microscope (**b**) images: the most distant point from the ICR on the fixed part of the comb-drive, the ICR of the comb-drive and the center of the ROI.

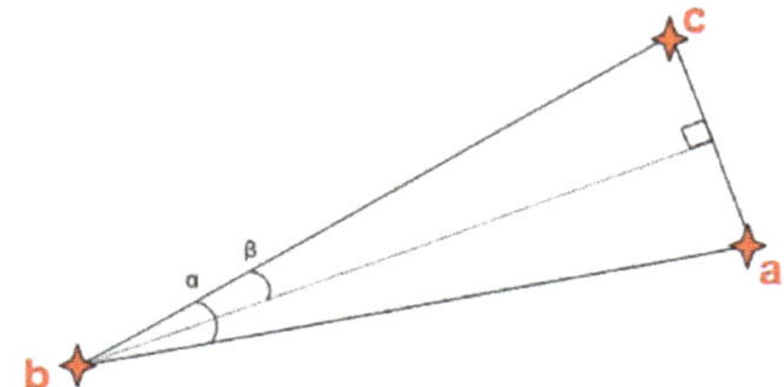

Figure 9. The triangle used by the software to evaluate the comb-drive angular displacement.

The above described software can be used with any type of image. For example, in this study two sets of images have been considered, acquired by the optical profilometer and the light microscope, respectively. The limit associated with this type of analysis is that the procedure provides a first semi-automatic part, which must be carried out by the operator. This approach introduces some sources of uncertainty, due to the variability in selecting the initial points for the determination of the comb-drive ICR, as well as the variability in the selection of the size and position of the Region of Interest, together with the uncertainty of the code itself. These sources of uncertainty will be considered in the next section, where a model will be proposed for the analysis of the uncertainty of both the entire measurement chains.

3.3. A Model for the Uncertainty Analysis of the Measurements

The purpose of this section is to estimate the overall uncertainty associated with the two measurement systems and for this reason a model for measurement uncertainty analysis is proposed; an analysis of the main uncertainty sources of the two measurement setups is carried out and the calculation of the measurement uncertainty relative to the angular displacements of the comb-drive device is performed, both for the profilometer and the light microscope.

For each value of the angular displacements, the mean $\bar{x}$ and standard deviation of the mean $S_{\bar{x}}$ have been obtained, based on a statistical analysis conducted on a number of observations $N = 12$ for the profilometer and $N = 16$ for the light microscope, for each considered voltage. On the hypothesis that the sample comes from a normal population, a Student's t-distribution has been used. For the calculation of the overall uncertainty of the measurement systems considered, it is necessary to combine the type A and type B uncertainties [47], using the following expression:

$$\delta_T = \sqrt{\delta_A^2 + \delta_B^2}\,.$$

(2)

First, the main uncertainty sources for both measurement systems have been identified and evaluated, as shown in Table 3, where for each source a probability density function *PDF* together with the uncertainty type and mean m is determined.

Table 3. OPS and LTM measurement setup uncertainty sources.

Uncertainty Source	Probability Density Function (*PDF*) [1]	Type [2]	m	δ (OPS)	δ (LTM)	Unit
Power Supply: voltage accuracy	$N(m,\sigma)$	B		0.06	0.04	V
Optical System: maximal lateral resolution due to diffraction	$U(m,\sigma)$	B	0	0.57	0.57	μm
Digital Image: digital conversion	$U(m,\sigma)$	B	0	3.14	1.43	μm
Image Processing Software: uncertainty in point identification	$U(m,\sigma)$	A		±1.9	±1.9	pixel
Uncertainty in ROI position (template-matching [3])				±1.9	±1.9	pixel
Uncertainty in ROI size (template-matching [3])				±1.9	±1.9	pixel
Uncertainty in the template-matching algorithm	Monte Carlo Simulation	B		0.02	0.02	°

Notes: [1] $N(\mu,\sigma)$ is a Gaussian *PDF* with mean m and standard deviation σ; $U(\mu,\sigma)$ is a uniform *PDF* with mean m and standard deviation σ; [2] Type A and type B uncertainty as in standard [47]; [3] the uncertainty value δ for OPS and LTM is referred to 95% of confidence level.

3.3.1. Uncertainty Analysis for OPS Measurement Setup

Type A, namely δ_A, uncertainty are evaluated by statistical methods (statistical analysis of a series of observations) and have been calculated from standard deviation of the experimental measurements; Type B, δ_B, uncertainty are evaluated by means other than the statistical analysis of series of observations. The main sources of type B uncertainty considered are:

- Power Supply uncertainty has been evaluated through the data sheet and a Gaussian distribution has been assumed;
- OPS uncertainty has been evaluated considering its maximal lateral resolution, limited by diffraction of the light; a uniform distribution with 0.6 μm semi-amplitude has been assumed;
- Digital image uncertainty has been evaluated considering the digital resolution of the acquired images (0.6 px/μm). A uniform distribution with 2 pixels semi-amplitude has been assumed [44,48,49];

3.3.2. Uncertainty Analysis for LTM Measurement Setup

In this case, δ_A still represents Type A uncertainty, whereas δ_B are the Type B uncertainty. The main sources of type B uncertainty considered are:

- Power Supply uncertainty has been evaluated through the data sheet and a Gaussian distribution has been assumed;
- LTM system uncertainty has been evaluated considering its maximal lateral resolution, limited by diffraction of the light; a uniform distribution with 0.6 μm semi-amplitude has been assumed.
- Digital image uncertainty has been evaluated considering the digital resolution of the acquired images (1.359 px/μm). A uniform distribution with 2 pixels semi-amplitude has been assumed.
- Uncertainty due to the plane planarity, the focusing plane variation, and the vibration of the DUT have been considered negligible, thanks to optical table with pneumatic vibration isolators that increase the stability of the device and reduce vibrations that could cause an incorrect characterization.

To evaluate the total uncertainty of the two measurement systems, according to Equation (2), two contributions have been determined: the first is related to the uncertainty contribution evaluable with a statistic analysis of the measurements dispersion obtained with the measurement system (δ_A), while the other is due to the overall uncertainty from the main error sources related to the experimental setup (δ_B). In order to evaluate the δ_B uncertainty, two main contributions have been determined: the first is associated to the derivative of the function that expresses the angular displacement θ of the comb-drive with respect to the applied voltage $\frac{\partial\theta}{\partial V}$, multiplied by the uncertainty due to the power supply δ_V, while the other is related to the measurement of the comb-drive angle $\delta\alpha_t$ [47]

$$\delta_B = \delta_\alpha = \sqrt{\left(\frac{\partial\theta}{\partial V}\delta_V\right)^2 + (\delta\alpha_t)^2}$$

(3)

In particular, a second order polynomial function

$$\vartheta = a \cdot V^2 + b \cdot V + c,$$

(4)

where a, b, and c are obtained experimentally, approximates the trend of the angular displacements.

The term $\delta\alpha_t$ is composed by two terms, the first $\delta\alpha_p$ is the angle measurement uncertainty due to the OPS and LTM system, associated to the variability of the parameters related to the manual measurement of the lengths Δx_a, Δx_b, Δy_a, and Δy_b, carried out by the operator; while the other $\delta\alpha_s$, is the angle measurement uncertainty due to the template-matching algorithm of image processing software. This last contribution is evaluated [44] by means of a Monte Carlo Simulation (with 10,000 iterations) implemented in MATLAB©; $\delta\alpha_s$ corresponds to $\pm 0.02°$, at 95% confidence level.

$$\delta\alpha_t = \sqrt{\left(\frac{\delta\alpha_p}{\alpha} \cdot \alpha\right)^2 + (\delta\alpha_s)^2}$$

(5)

The angle measurement uncertainty $\delta\alpha_p$ has been measured by means of the triangular properties as shown in Figure 10. For the measurement of values Δx_a, Δx_b, Δy_a and Δy_b, 10 tests were carried out for each of the segments and their mean value has been considered.

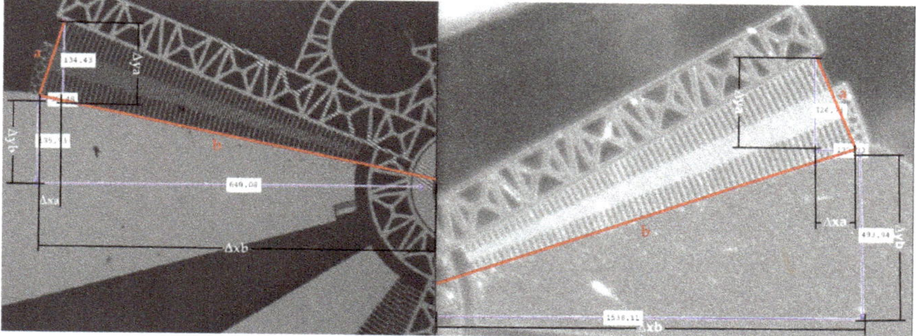

Figure 10. Angle measurement for the uncertainty evaluation related to the profilometer image, on the **left** and microscope image, on the **right**.

The uncertainty related to the light microscope and profilometer error can be obtained as approximated in [44]:

$$\frac{\delta\alpha_p}{\alpha} = \sqrt{\left(\frac{\delta a}{a}\right)^2 + \left(\frac{\delta b}{b}\right)^2},$$

(6)

where

$$a = \sqrt{\Delta x_a{}^2 + \Delta y_a{}^2}$$
$$b = \sqrt{\Delta x_b{}^2 + \Delta y_b{}^2}$$

$$(7)$$

and

$$\delta a = \sqrt{\left(\frac{\partial a}{\partial x}\delta x\right)^2 + \left(\frac{\partial a}{\partial y}\delta y\right)^2} = \sqrt{\left(\frac{\Delta x_a}{\sqrt{\Delta x_a^2 + \Delta y_a^2}}\delta x\right)^2 + \left(\frac{\Delta y_a}{\sqrt{\Delta x_a^2 + \Delta y_a^2}}\delta y\right)^2}$$

$$\delta b = \sqrt{\left(\frac{\partial b}{\partial x}\delta x\right)^2 + \left(\frac{\partial b}{\partial y}\delta y\right)^2} = \sqrt{\left(\frac{\Delta x_b}{\sqrt{\Delta x_b^2 + \Delta y_b^2}}\delta x\right)^2 + \left(\frac{\Delta y_b}{\sqrt{\Delta x_b^2 + \Delta y_b^2}}\delta y\right)^2}$$

$$(8)$$

The quantities a and b in (7) depend on the image resolution and size. For this study, the lengths of the comb-drive triangle in the profilometer image are considered for the maximum rotation ($1.3°$), i.e., $\Delta x_a = 25$ pixel, $\Delta y_a = 150$ pixel, $\Delta x_b = 680$ pixel, and $\Delta y_b = 150$ pixel, therefore $a = 152$ pixel, and $b = 696$ pixel. Instead, the lengths of the comb-drive triangle in the microscope image are considered for the maximum rotation ($0.93°$), i.e., $\Delta x_a = 130$ pixel, $\Delta y_a = 330$ pixel, $\Delta x_b = 1550$ pixel, and $\Delta y_b = 450$ pixel, therefore $a = 355$ pixel, and $b = 1614$ pixel.

4. A Comparison between the Two Measurement Systems

At first, a comparison between the results obtained from the uncertainty analysis for both image acquisition systems has been carried out; from the results obtained through the analysis software it has been possible to collect a series of measurements attributable to the comb-drive angular displacement depending on the voltage supply. In particular, 12 measurements were obtained for each applied voltage value, for the OPS, and 16 measurements for each applied voltage value, for the LTM system. In a second stage, it has been necessary to verify whether the two sets of results are comparable, within the interval of the experimental uncertainties.

To verify the initial hypothesis that, using two different measurement systems, the angular displacement measurement of the comb-drive is the same, the approach described in Ref. [50] has been followed. Considering the average values, the total uncertainties δ_T of OPS and LTM have been obtained by using Equation (2) and then reported in the form of standard deviations in Table 4. The available data are expressed in the form:

$$\hat{X} = \bar{X} \pm \delta_X \tag{9}$$
$$\hat{Y} = \bar{Y} \pm \delta_Y \tag{10}$$

where $\bar{X}$ and $\bar{Y}$ are the mean values of X and Y (OPS and LTM measurements, respectively), while $\hat{X}$ and $\hat{Y}$ represent the measured values according to the standard [47]. To evaluate whether the two different measurements can be considered consistent or not, it is necessary to find the best estimate for the difference $\Delta_{XY} = \hat{X} - \hat{Y}$ and establish the highest and the lowest probable values of Δ_{XY}. The highest probable value for Δ_{XY} is obtained if $\bar{X}$ assumes its higher probable value, $\bar{X} + \delta_X$, and at the same time $\hat{Y}$ assumes its lowest probable value, $\bar{Y} - \delta_Y$. In this way the highest probable value for Δ_{XY} is

$$p_{max} = (\bar{X} + \delta_X) - (\bar{Y} - \delta_Y) = (\bar{X} - \bar{Y}) + (\delta_X + \delta_Y) \ . \tag{11}$$

Similarly, the lowest probable value for Δ_{XY} is obtained if $\hat{X}$ assumes its lowest probable value, $\bar{X} - \delta_X$, and at the same time $\hat{Y}$ assumes its highest probable value, $\bar{Y} + \delta_Y$,

$$p_{max} = (\bar{X} - \delta_X) - (\bar{Y} + \delta_Y) = (\bar{X} - \bar{Y}) - (\delta_X + \delta_Y) \ . \tag{12}$$

Combining (11) and (12), the difference between the measured values is

$$\Delta_{XY} = (\bar{X} - \bar{Y}) \pm (\delta_X + \delta_Y) \, . \tag{13}$$

The evaluation of the difference $\left|\bar{X} - \bar{Y}\right|$ and the sum $\delta_X + \delta_Y$ is fundamental for the sake of our investigation. In fact, if the difference $\left|\bar{X} - \bar{Y}\right|$ has the same order of magnitude as, or even less than, the sum $\left|\bar{X} - \bar{Y}\right|$, then the two different systems, namely OPS and LTM, measure the angular displacement of the comb-drive without significant difference.

5. Results

All the measurement data have been processed and interpolated to provide a curve fitting of the motion of the comb-drive depending on the voltage supply; in Figure 11 the two curves related to the two measurement setups are shown, while the detailed results are reported in Tables 4.

Table 4. Angular rotation, Total Uncertainty, expressed as standard deviation, and Total Relative Uncertainty of comb-drive depending on the applied voltage.

Supply Voltage (V)	Angular Displ. Mean Value (°) (OPS)	Angular Displ. Mean Value (°) (LTM)	Total Uncertainty (°) (OPS)	Total Uncertainty (°) (LTM)	Total Relative Uncertainty (%) (OPS)	Total Relative Uncertainty (%) (LTM)
2	0.005	0.011	0.010	0.019	206.5%	170.7%
4	0.003	0.023	0.013	0.023	460.1%	99.0%
6	0.065	0.060	0.011	0.023	16.7%	39.2%
8	0.110	0.108	0.015	0.028	13.8%	25.6%
10	0.160	0.168	0.010	0.025	6.5%	14.8%
12	0.227	0.237	0.011	0.028	4.8%	11.8%
14	0.319	0.321	0.011	0.025	3.5%	7.9%
16	0.427	0.414	0.017	0.030	4.0%	7.2%
18	0.543	0.533	0.013	0.034	2.3%	6.4%
20	0.696	0.642	0.016	0.024	2.2%	3.7%
22	0.810	0.787	0.017	0.019	2.1%	2.5%
24	0.952	0.926	0.021	0.030	2.2%	3.2%
26	1.122		0.023		2.0%	
28	1.308		0.021		1.6%	

For low values of the voltage, the observations reveal that the microgripper mobile parts face a certain resistance while attempting to move, showing a certain instability. In fact, the torque-voltage function presents an increasing rate of change. This phenomenon is one of the main source of a high dispersion of the results at low values of voltage, while this problem does not occur for higher values (>4 V).

In Table 5 the results of the comparison between the OPS and LTM systems are shown.

As already observed in Ref. [44], for the optical profilometer, even if we consider the image acquisition system of the light microscope, it is possible to observe a concordant behavior with the results reported in Refs. [35,37].

Second order polynomial regression curves have been determined by a least squares fitting method to describe the outcome from the OPS and the LTM systems: the fitting capacity is confirmed by the high value R^2 coefficients [51], equal to 0.999 for both the OPS and LTM systems.

The Equations (3) and (4) have been used to calculate the type B uncertainty δ_B.

Table 5. Applied voltage vs parameters differences.

Applied Voltage (V)	$\bar{X} \pm \delta_{T_X}$ [°]	$\bar{Y} \pm \delta_{T_Y}$ [°]	$\lvert \bar{X}_V - \bar{Y}_V \rvert$ [°]	$\delta_{T_X} + \delta_{T_Y}$ [°]
2	0.005 ± 0.010	0.011 ± 0.019	0.006	0.029
4	0.003 ± 0.013	0.023 ± 0.023	0.02	0.04
6	0.065 ± 0.011	0.060 ± 0.023	0.005	0.03
8	0.110 ± 0.015	0.108 ± 0.028	0.002	0.04
10	0.160 ± 0.010	0.168 ± 0.025	0.008	0.04
12	0.227 ± 0.011	0.237 ± 0.028	0.01	0.04
14	0.319 ± 0.011	0.321 ± 0.025	0.002	0.04
16	0.427 ± 0.017	0.414 ± 0.03	0.013	0.05
18	0.543 ± 0.013	0.533 ± 0.03	0.01	0.05
20	0.696 ± 0.016	0.642 ± 0.024	0.054	0.04
22	0.810 ± 0.017	0.787 ± 0.019	0.023	0.04
24	0.952 ± 0.021	0.926 ± 0.03	0.026	0.05

Table 4 shows that above 16 V the total relative uncertainty of the optical profilometer is less than 4.0%, while for the microscope it is less than 7.2%. Since the same software for digital image processing has been used, these results may be mainly related to the two different setups. The greatest contribution in the calculation of uncertainty is given by the δ_{α_t} uncertainty, composed by the angle measurement uncertainty due to the OPS and LTM, $\left(\frac{\delta_{\alpha p}}{\alpha} \cdot \alpha \right)$, and the angle measurement uncertainty due to the template-matching algorithm of the image processing software δ_{α_s}, that correspond to 0.02° both for OPS and LTM systems. As mentioned above, the OPS and the LTM generally have different typical values of SNR and this fact is also affecting uncertainty in the two measurement systems.

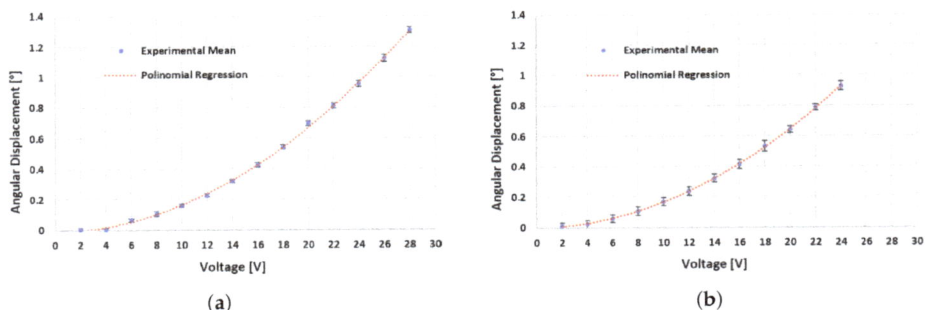

Figure 11. Relationship between angular displacement vs the applied voltage for the OPS (**a**) and the LTM (**b**) systems.

A second evaluation of the two experimental setups has been conducted; indeed, it has been necessary to evaluate whether the different results relative to the mean values of the angular displacement of the comb drive for the two experimental setups are comparable or not, to properly establish if the two measurement systems produce the same results, based on the verification that the differences between the measured mean values are smaller or comparable with the sum of the original uncertainties. Table 5 shows that, for each applied voltage value, there is no significant difference between the results obtained by the two methods. It is therefore possible to conclude that the measurements carried out with the OPS and LTM systems can be considered consistent, within the interval of the experimental uncertainties. From the point of view of total relative uncertainty, the results obtained with the profilometer appear to be better than those obtained with the microscope. However, considering the high costs of a profilometer system (about two times higher than the light microscope here used) and also the ease of use of a system such as the light microscope, the latter appears to be the best trade-off system able to carry out performance characterization of a device like microgripper for biomedical application. Furthermore, despite both image acquisition systems being able to characterize the microgripper in operating conditions, only the microscope permits a real-time study; this last feature is certainly the most important because it allows LTM to characterize the device in static and dynamic conditions.

6. Discussion

The first step of this investigation has been the mechanical characterization of a microgripper prototype. The comb-drive angular displacement has been expressed as a function of the applied voltage. Two different measurement systems, OPS and LTM, have been used. Both systems showed that the two rotation-voltage curves follow the same quadratic trend. In order to perform the measurements, a systematic approach has been developed based on in-house built software. This gave rise to a system which has quite high repeatability and low-operator-dependence. Moreover, an analysis of the uncertainties has been carried out, with the construction of a model that considers the main sources of uncertainty present in the measuring setups. The evaluation of the accuracy of the considered setups has shown that for voltages greater than 14 V, the total relative uncertainty of OPS is less than 4.0%, while for the LTM is less than 7.2%. A verification to check whether the two sets of results are consistent, within the interval of the experimental uncertainties has been carried out.

The same in-house software has been used for image processing and therefore the differences in the obtained results are due to the uncertainty of the two image acquisition systems. The main considered contributions were:

- Type A uncertainty δ_A evaluated by means of a statistic analysis of the measurements results;
- the angle measurement uncertainty δ_{α_p} due to the OPS and LTM systems;
- the angle measurement uncertainty δ_{α_s} due to the template-matching algorithm of image processing software and
- the power supply uncertainty δ_V.

Despite the obtained results showing that the OPS system has a total relative uncertainty lower than those of the LTM system, the light microscope is still the image acquisition system that will best suit the characterization of MEMS-Technology based microgrippers. In fact, profilometer or electronic microscopy are not practical or even unfeasible for the characterization of these devices in working or in real-time conditions. Therefore, the light microscope is the most promising image acquisition system because it may perform a mechanical characterization, and because of its use in biomedical or surgical real-time scenarios.

7. Conclusions

Nowadays there is a growing need of more accurate and precise devices, especially in microsurgery or diagnostics. The developments of the manufacturing technology offer new tools, such as microgrippers, suitable for a variety of biomedical applications, including minimally invasive surgery. However, SEM observation is impossible in air, wet or liquid environments, while profilometer or OCT measurements are often unpractical or more expensive in operational or real-time scenarios. The aim of this paper is therefore encouraging the use of light microscopes in future investigations, by validating, for the test case, light microscopy as the best trade-off among the characterization accuracy and the need of a large operational range and real time response. The validation method has been applied to a MEMS-technology based microgripper developed earlier by the research group.

Author Contributions: Conceptualization and investigation, all the Authors; methodology, A.S.; software, data curation and validation, F.V. and F.O.; formal analysis, F.V., F.O. and A.S.; writing, N.P.B.; supervision, S.A.S. and P.U.; resources, S.A.S. and R.C.

Funding: This research received no external funding.

Acknowledgments: The microgrippers under study have been fabricated by MTLAB-FBK, Micro Technology Laboratory-Fondazione Bruno Kessler.

Conflicts of Interest: The authors declare no conflict of interest.

Abbreviations

The following abbreviations are used in this manuscript:

SEM   Scanning Electron Microscope
MEMS   Micro Electro Mechanical System
IAS   Image Analysis Software
OPS   Optical Profilometer System
LTM   Light Trinocular Microscope

References

1. Rebello, K.J. Applications of MEMS in surgery. *Proc. IEEE* **2004**, *92*, 43–55. [CrossRef]
2. Verotti, M.; Dochshanov, A.; Belfiore, N. A Comprehensive Survey on Microgrippers Design: Mechanical Structure. *J. Mech. Des. Trans. ASME* **2017**, *139*, 060801. [CrossRef]
3. Dochshanov, A.; Verotti, M.; Belfiore, N. A Comprehensive Survey on Microgrippers Design: Operational Strategy. *J. Mech. Des. Trans. ASME* **2017**, *139*, 070801. [CrossRef]
4. Lowrie, C.; Desmulliez, M.P.Y.; Hoff, L.; Elle, O.J.; Fosse, E. Fabrication of a MEMS accelerometer to detect heart bypass surgery complications. *Sens. Rev.* **2009**, *29*, 319–325. [CrossRef]
5. Balla, A.; Quaresima, S.; Ursi, P.; Seitaj, A.; Palmieri, L.; Badiali, D.; Paganini, A.M. Hiatoplasty with crura buttressing versus hiatoplasty alone during laparoscopic sleeve gastrectomy. *Gastroenterol. Res. Pract.* **2017**, *2017*, 6565403. [CrossRef]
6. Popivanov, G.; Tabakov, M.; Mantese, G.; Cirocchi, R.; Piccinini, I.; D'Andrea, V.; Covarelli, P.; Boselli, C.; Barberini, F.; Tabola, R.; Ursi, P.; Cavaliere, D.; Surgical treatment of gastrointestinal stromal tumors of the duodenum: A literature review. *Transl. Gastroenterol. Hepatol.* **2018**, *3*, 71. [CrossRef] [PubMed]
7. Cochetti, G.; Del Zingaro, M.; Boni, A.; Cocca, D.; Panciarola, M.; Tiezzi, A.; Gaudio, G.; Balzarini, F.; Ursi, P.; Mearini, E. Colovesical fistula: Review on conservative management, surgical techniques and minimally invasive approaches. *G. Chir.* **2018**, *39*, 195–207.
8. Quaresima, S.; Balla, A.; Dambrosio, G.; Bruzzone, P.; Ursi, P.; Lezoche, E.; Paganini, A.M. Endoluminal loco-regional resection by TEM after R1 endoscopic removal or recurrence of rectal tumors. *Minim. Invasive Ther. Allied Technol.* **2016**, *25*, 134–140. [CrossRef] [PubMed]
9. Paci, M.; Scoglio, D.; Ursi, P.; Barchetti, L.; Fabiani, B.; Ascoli, G.; Lezoche, G. Transanal Endocopic Microsurgery (TEM) in advanced rectal cancer disease treatment [Il ruolo della TEM nel trattamento dei tumori del retto extraperitoneale]. *Ann. Ital. Chir.* **2010**, *81*, 269–274.
10. Lezoche, E.; Fabiani, B.; D'Ambrosio, G.; Ursi, P.; Balla, A.; Lezoche, G.; Monteleone, F.; Paganini, A.M. Nucleotide-guided mesorectal excision combined with endoluminal locoregional resection by transanal endoscopic microsurgery in the treatment of rectal tumors: Technique and preliminary results. *Surg. Endosc.* **2013**, *27*, 4136–4141. [CrossRef] [PubMed]
11. Ursi, P.; Santoro, A.; Gemini, A.; Arezzo, A.; Pironi, D.; Renzi, C.; Cirocchi, R.; Di Matteo, F.; Maturo, A.; D'Andrea, V.; et al. Comparison of outcomes following intersphincteric resection vs low anterior resection for low rectal cancer: A systematic review. *G. Chir.* **2018**, *39*, 123–142.
12. Theillet, P.O.; Pierron, O.N. Quantifying adsorbed water monolayers on silicon MEMS resonators exposed to humid environments. *Sens. Actuators A Phys.* **2011**, *171*, 375–380. [CrossRef]
13. Bhattacharya, S.; Murali Krishna, A.; Lombardi, D.; Crewe, A.; Alexander, N. Economic MEMS based 3-axis water proof accelerometer for dynamic geo-engineering applications. *Soil Dyn. Earthq. Eng.* **2012**, *36*, 111–118. [CrossRef]
14. Neagu, C.; Jansen, H.; Gardeniers, H.; Elwenspoek, M. Electrolysis of water: An actuation principle for MEMS with a big opportunity. *Mechatronics* **2000**, *10*, 571–581. [CrossRef]
15. Scherge, M.; Li, X.; Schaefer, J.A. The effect of water on friction of MEMS. *Tribol. Lett.* **1999**, *6*, 215–220. [CrossRef]

16. Yabuki, T.; Nakabeppu, O. Heat transfer mechanisms in isolated bubble boiling of water observed with MEMS sensor. *Int. J. Heat Mass Transf.* **2014**, *76*, 286–297. [CrossRef]

17. Yao, J.; Huang, C.H.; Wang, L.; Yang, J.; Gao, L.; Maslov, K.I.; Zou, J.; Wang, L.V. Wide-field fast-scanning photoacoustic microscopy based on a water-immersible MEMS scanning mirror. *J. Biomed. Opt.* **2012**, *17*, 080505. [CrossRef]

18. Bashir, R. BioMEMS: State-of-the-art in detection, opportunities and prospects. *Adv. Drug Deliv. Rev.* **2004**, *56*, 1565–1586. [CrossRef]

19. Bhushan, B. Nanotribology and nanomechanics of MEMS/NEMS and BioMEMS/BioNEMS materials and devices. *Microelectron. Eng.* **2007**, *84*, 387–412. [CrossRef]

20. Grayson, A.; Shawgo, R.; Johnson, A.; Flynn, N.; Li, Y.; Cima, M.; Langer, R. A BioMEMS review: MEMS technology for physiologically integrated devices. *Proc. IEEE* **2004**, *92*, 6–21. [CrossRef]

21. Bragheri, F.; Minzioni, P.; Martinez Vazquez, R.; Bellini, N.; Paiè, P.; Mondello, C.; Ramponi, R.; Cristiani, I.; Osellame, R. Optofluidic integrated cell sorter fabricated by femtosecond lasers. *Lab Chip* **2012**, *12*, 3779–3784. [CrossRef]

22. Wu, H.; Kuhn, T.; Moy, V. Mechanical properties of L929 cells measured by atomic force microscopy: Effects of anticytoskeletal drugs and membrane crosslinking. *Scanning* **1998**, *20*, 389–397. [CrossRef]

23. Mattana, S.; Mattarelli, M.; Urbanelli, L.; Sagini, K.; Emiliani, C.; Serra, M.; Fioretto, D.; Caponi, S. Non-contact mechanical and chemical analysis of single living cells by microspectroscopic techniques. *Light Sci. Appl.* **2018**, *7*, 17139. [CrossRef]

24. Mulvany, M.J.; Halpern, W. Mechanical properties of vascular smooth muscle cells in situ. *Nature* **1976**, *260*, 617–619. [CrossRef]

25. Wakatsuki, T.; Schwab, B.; Thompson, N.; Elson, E. Effects of cytochalasin D and latrunculin B on mechanical properties of cells. *J. Cell Sci.* **2001**, *114*, 1025–1036.

26. Levesque, P.; Gauvin, R.; Larouche, D.; Auger, F.; Germain, L. A Computer-Controlled Apparatus for the Characterization of Mechanical and Viscoelastic Properties of Tissue-Engineered Vascular Constructs. *Cardiovasc. Eng. Technol.* **2011**, *2*, 24–34. [CrossRef]

27. Bagolini, A.; Bellutti, P.; Di Giamberardino, P.; Rudas, I.J.; D'Andrea, V.; Verotti, M.; Dochshanov, A.; Belfiore, N.P. Stiffness Characterization of Biological Tissues by Means of MEMS-Technology Based Micro Grippers Under Position Control. In *Advances in Service and Industrial Robotics*; Ferraresi, C., Quaglia, G., Eds.; Springer International Publishing: Cham, Switzerland, 2018; pp. 939–947.

28. Belfiore, N.; Broggiato, G.; Verotti, M.; Balucani, M.; Crescenzi, R.; Bagolini, A.; Bellutti, P.; Boscardin, M. Simulation and construction of a mems CSFH based microgripper. *Int. J. Mech. Control* **2015**, *16*, 21–30.

29. Verotti, M.; Crescenzi, R.; Balucani, M.; Belfiore, N.P. MEMS-based conjugate surfaces flexure hinge. *J. Mech. Des. Trans. ASME* **2015**, *137*, 012301. [CrossRef]

30. Botta, F.; Verotti, M.; Bagolini, A.; Bellutti, P.; Belfiore, N.P. Mechanical response of four-bar linkage microgrippers with bidirectional electrostatic actuation. *Actuators* **2018**, *7*, 78. [CrossRef]

31. Cecchi, R.; Verotti, M.; Capata, R.; Dochshanov, A.; Broggiato, G.; Crescenzi, R.; Balucani, M.; Natali, S.; Razzano, G.; Lucchese, F.; et al. Development of micro-grippers for tissue and cell manipulation with direct morphological comparison. *Micromachines* **2015**, *6*, 1710–1728. [CrossRef]

32. Di Giamberardino, P.; Bagolini, A.; Bellutti, P.; Rudas, I.; Verotti, M.; Botta, F.; Belfiore, N.P. New MEMS tweezers for the viscoelastic characterization of soft materials at the microscale. *Micromachines* **2017**, *9*, 15. [CrossRef] [PubMed]

33. Verotti, M.; Dochshanov, A.; Belfiore, N.P. Compliance Synthesis of CSFH MEMS-Based Microgrippers. *J. Mech. Des. Trans. ASME* **2017**, *139*, 022301. [CrossRef]

34. Potrich, C.; Lunelli, L.; Bagolini, A.; Bellutti, P.; Pederzolli, C.; Verotti, M.; Belfiore, N.P. Innovative silicon microgrippers for biomedical applications: Design, mechanical simulation and evaluation of protein fouling. *Actuators* **2018**, *7*, 12. [CrossRef]

35. Bagolini, A.; Ronchin, S.; Bellutti, P.; Chistè, M.; Verotti, M.; Belfiore, N.P. Fabrication of Novel MEMS Microgrippers by Deep Reactive Ion Etching With Metal Hard Mask. *J. Microelectromech. Syst.* **2017**, *26*, 926–934. [CrossRef]

36. Bertini, S.; Verotti, M.; Bagolini, A.; Bellutti, P.; Ruta, G.; Belfiore, N.P. Scalloping and stress concentration in DRIE-manufactured comb-drives. *Actuators* **2018**, *7*, 57. [CrossRef]

37. Crescenzi, R.; Balucani, M.; Belfiore, N. Operational characterization of CSFH MEMS technology based hinges. *J. Micromech. Microeng.* **2018**, *28*, 5. [CrossRef]
38. Prasad, P. *Introduction to Biophotonics*; John Wiley and Sons, Inc.: Hoboken, NJ, USA, 2003.
39. Strathman, M.; Liu, Y.; Keeler, E.G.; Song, M.; Baran, U.; Xi, J.; Sun, M.T.; Wang, R.; Li, X.; Lin, L.Y. MEMS scanning micromirror for optical coherence tomography. *Biomed. Opt. Express* **2015**, *6*, 211–224. [CrossRef]
40. Werkmeister, R.M.; Sapeta, S.; Schmidl, D.; Garhöfer, G.; Schmidinger, G.; dos Santos, V.A.; Aschinger, G.C.; Baumgartner, I.; Pircher, N.; Schwarzhans, F.; et al. Ultrahigh-resolution OCT imaging of the human cornea. *Biomed. Opt. Express* **2017**, *8*, 1221–1239. [CrossRef]
41. Huang, D.; Swanson, E.A.; Lin, C.P.; Schuman, J.S.; Stinson, W.G.; Chang, W.; Hee, M.R.; Flotte, T.; Gregory, K.; Puliafito, C.A. Optical Coherence Tomography. *Science* **1991**, *254*, 1178–1181. [CrossRef]
42. Lee, J.; Shirazi, M.F.; Park, K.; Jeon, M.; Kim, J. Defect inspection of actuator lenses using swept-source optical coherence tomography. *Opt. Rev.* **2018**, *25*, 403–409. [CrossRef]
43. Kim, S.; Lee, C.; Kim, J.Y.; Kim, J.; Lim, G.; Kim, C. Two-axis polydimethylsiloxane-based electromagnetic microelectromechanical system scanning mirror for optical coherence tomography. *J. Biomed. Opt.* **2016**, *21*, 106001. [CrossRef]
44. Orsini, F.; Vurchio, F.; Scorza, A.; Crescenzi, R.; Sciuto, S.A. An image analysis approach to microgrippers displacement measurement and testing. *Actuators* **2018**, *7*, 64. [CrossRef]
45. Chang, R.J.; Chien, Y.C. Polymer microgripper with autofocusing and visual tracking operations to grip particle moving in liquid. *Actuators* **2018**, *7*, 27. [CrossRef]
46. Brown, L. Imaging Particle Analysis: Resolution and Sampling Considerations. 2009. Accessed online: http://www.fluidimaging.com/resource-center-whitepapers (accessed on 12 June 2015).
47. International Organization for Standardization; International Electrotechnical Commission. *ISO/IEC Guide 98-3: Uncertainty of Measurement. Part 3: Guide to the Expression of Uncertainty in Measurement (GUM:1995). Partie 3: Guide pour L'expression de L'incertitude de Mesure (GUM:1995)*; Number pt. 3; ISO Standard Catalogue: Geneva, Switzerland, 2008.
48. Pawley, J. *Handbook of Biological Confocal Microscopy*; Springer US: New York, NY, USA, 2010.
49. Waters, J. Accuracy and precision in quantitative fluorescence microscopy. *J. Cell Biol.* **2009**, *185*, 1135–1148. [CrossRef] [PubMed]
50. Taylor, J.R. *An Introduction to Error Analysis. Uncertainty Stuty in Physical Measurements. (in Italian) Introduzione all'analisi degli Errori. Lo Studio delle Incertezze Nelle Misure Fisiche*; Collana di Fisica. Testi e Manuali; Zanichelli: Bologna, Italy, 1986.
51. Navidi, W. *Statistics for Engineers and Scientists*; McGraw-Hill: New York, NY, USA, 2014.

MDPI

St. Alban-Anlage 66

4052 Basel

Switzerland

Tel. +41 61 683 77 34

Fax +41 61 302 89 18

www.mdpi.com

Applied Sciences Editorial Office

E-mail: applsci@mdpi.com

www.mdpi.com/journal/applsci

www.ingramcontent.com/pod-product-compliance
Lightning Source LLC
La Vergne TN
LVHW071005170726
843515LV00006B/1083